## Science and Christianity

Science and Christianity

Foundations and Frameworks for Moving Forward in Faith

TIM REDDISH

WIPF & STOCK · Eugene, Oregon

SCIENCE AND CHRISTIANITY
Foundations and Frameworks for Moving Forward in Faith

Copyright © 2016 Tim Reddish. All rights reserved. Except for brief quotations in critical publications or reviews, no part of this book may be reproduced in any manner without prior written permission from the publisher. Write: Permissions, Wipf and Stock Publishers, 199 W. 8th Ave., Suite 3, Eugene, OR 97401.

Wipf & Stock
An Imprint of Wipf and Stock Publishers
199 W. 8th Ave., Suite 3
Eugene, OR 97401

www.wipfandstock.com

PAPERBACK ISBN: 978-1-4982-9604-5
HARDCOVER ISBN: 978-1-4982-9606-9
EBOOK ISBN: 978-1-4982-9605-2

Manufactured in the U.S.A.   08/09/16

All scriptural references are from the NRSV Bible translation.

Excerpt from John Donne's *An Anatomy of the World* (1611) is in the public domain. (See chapter 1.)

Figure 1 in chapter 1; by Mary Reddish. Used by permission.

Figure 1 in chapter 3 contains two images, both of which are in the public domain; details below:
"My Wife and My Mother-in-Law" by W.E. Hill 1915. https://commons.wikimedia.org/wiki/File:My_Wife_and_My_Mother-In-Law_%28Hill%29.svg.

Duck and rabbit optical illusion, "The Mind's Eye," Joseph Jastrow (1899), from *Popular Science Monthly* 54. https://commons.wikimedia.org/wiki/File:PSM_V54_D328_Optical_illusion_of_a_duck_or_a_rabbit_head.png.

For my family, especially my wife, Mary.

Contents

List of Illustrations and Tables | viii
Acknowledgments | ix
Introduction | xi

Chapter 1
Science and Scripture: The Bible on Trial | 1

Chapter 2
On the Inspiration and Interpretation of Scripture | 23

Chapter 3
On the Nature of Science | 42

Chapter 4
On Ways of Relating Science and Christianity | 60

Chapter 5
On Chance, Order, and Necessity | 92

Chapter 6:
On the Nature of God | 109

Chapter 7
On Miracles and Prayer | 125

Chapter 8
Revisiting Science and Scripture: Creation Texts
in the Old Testament | 144

Afterword | 169

Appendix 1
Traditional Theistic Arguments for the Existence of God | 171

Appendix 2
A Brief Excursion into Metaphysics | 175

Bibliography | 179
General Index | 187

Illustrations and Tables

Figure 1: Ancient Hebrew Conception of the World | 22

Figure 1a: Duck and Rabbit Optical Illusion: "The Mind's Eye," by Joseph Jastrow (1899) | 48

Figure 1b: "My Wife and My Mother-in-Law," by W. E. Hill (1915) | 48

Table 1: A Framework of God's Creative Activity in Genesis 1 | 148

Acknowledgments

Many years ago I wrote a draft "book" on this topic based on my thinking at the time. I showed a copy to my dad, who studied theology at St. Andrews University in Scotland. He encouraged me to get the book published. Much of my thinking has changed since then—for the better! My dad died in 2013, but I know that he would have been thrilled with the final outcome.

I would also like to thank Andrew and Mary Templer, who have closely followed all aspects of my journey since coming to Canada in 2002. Thank you for encouraging me in my transition from physics to theology.

More recently, I have been a theology student at Knox College, Toronto. I express my thanks to my professors and fellow seminary students; thank you for all you have taught me.

As you read this book, it will be evident that certain authors have influenced my thinking in this area; I have cited their works throughout. I am particularly indebted to John Polkinghorne, Alister McGrath, Ian Barbour, Russell Stannard, David Wilkinson, and Keith Ward—whom I have recently discovered. And I acknowledge the many theologians who have shaped my thinking, including Clark Pinnock, Lesslie Newbigin, Terence Fretheim, and N. T. Wright.

I would also like to thank Prof. Bill Crosby for the opportunity to team-teach a course on *Science and Christianity* at Canterbury College, Windsor. I greatly enjoyed the experience and the chance to test drive, as it were, some of the material in this book.

I also acknowledge the fine work of the whole team at Wipf & Stock for bringing this project to fruition.

And finally, and—of course—by no means least, I thank my wife, Mary. Thank you for your encouragement as we share this adventure together. I love you.

Introduction

"What, *another* book on 'Science and Christianity'?" That was my first reaction when friends encouraged me to undertake such a project. So much has been written on this topic on both sides of the Atlantic that I was not sure what new contributions I could meaningfully add. Particularly as the excellent material by giants in the area of science-faith dialogue, such as John Polkinghorne, Ian Barbour, Alister McGrath, and Keith Ward, have had seemingly little effect in the life of the North American church. On reflection, I realized that there is a major disconnect between discussions in scholarly circles on this theme, which have been ongoing for many decades now, and the regular life of church congregations—particularly within some branches of the evangelical church.

To compound the problem, the established churches in the West have been in serious, steady numerical decline over the last fifty years. It is particularly evident in the young adult age range. In *You Lost Me: Why Young Christians Are Leaving Church . . . and Rethinking Faith*, David Kinnaman details, in six chapters, the major reasons that emerging adults (18–29 years old) are losing interest in our congregations. One chapter is simply titled "Antiscience."[1] This generation sees the church as standing against the findings of science. Yet over 50 percent of young people aspire to science-related careers like biology, chemistry, engineering, and technology, along with the medical and health-related professions. In light of that, how many pastors or youth workers had addressed issues of faith and science in the course of the year? One percent.[2] This reflects a serious discontinuity between our culture and faith. Or worse, if it indicates the church is in a state of denial.

---

1. Kinnaman, *You Lost Me*, 139–48.
2. Cootsona, "When Science Comes to Church," 1.

Introduction

The real issue is not about reversing declining church numbers, but in the way we *think*. In his acclaimed study, Mark Noll writes: "Evangelical thinking about science is still but a shadow of what God, nature, and the Christian faith deserve."[3] Noll recently revisited his prophetic plea concerning the health of the evangelical mind; the section on science concludes:

> Satisfactory resolution to problems stemming from responsible biblical interpretation brought together with responsible interpretation of nature will not come easily. Such resolution requires *more sophistication in scientific knowledge, more sophistication in biblical hermeneutics*, and more humility of spirit than most of us possess. But it is not wishful thinking to believe that such resolution is possible. It is rather an expected hope that goes directly from confidence in what has been revealed in Jesus Christ.[4]

There is *still* a need for many Christians to courageously engage in the findings of modern science, which can only begin if we are prepared to acknowledge our ignorance and fear in order to travel on a journey of discovery. There is also a desperate need for Christians to be more self-critical in the way we view Scripture—its inspiration and interpretation.[5] As we participate in this ongoing science and Christianity expedition, are we prepared to trust the Holy Spirit to guide us into all the truth on this faith adventure (John 16:13)? Are we prepared to *risk* being *changed* by the process? Or have we already firmly made up our minds? That is the challenge every person faces—those with faith and those who claim none.

In light of the ongoing crisis identified by Noll (and others) what is the church to do? One response is for church leaders and enquiring minds to be better informed on science and faith. It is only from a framework of critical engagement with both science and the Bible that contemporary issues and the needs of the church and society can be addressed. This is already underway with initiatives by various groups, such as the BioLogos Foundation, the American Scientific Affiliation—and their international partners,

---

3. Noll, *Scandal of the Evangelical Mind*, 233.

4. Noll, *Jesus Christ and the Life of the Mind*, 124, emphasis mine. He adds, "If, therefore, humbly responsible thinkers, properly equipped scientifically and hermeneutically, conclude that the full picture of human evolution now standard in many scientific disciplines fits with a trustworthy interpretation of Scripture, that conclusion can be regarded as fully compatible with historic Christian orthodoxy as defined by the normative creeds." Ibid., 124.

5. This topic is explored in Enns, *The Bible Tells Me So*, and Enns, *Inspiration and Incarnation*.

INTRODUCTION

the Faraday Institute of Science and Religion, and many others.[6] Yet much more evidently needs to be done. Consequently, my goal here is to communicate to ministers and seminary students. Better informed church leaders can then engage their congregations and so foster transformational faith. It is also for jaded, thoughtful Christians who wrestle to maintain their faith in a church climate that is intellectually unsatisfying and stifling. The book is also for the agnostic and the simply curious who are sympathetic to Christianity but deeply suspicious of institutional religion.

This book brings material from various sources together and presents them in—hopefully—an accessible and engaging way, especially for those who do not have a science background. Even so, I freely admit the book is not a light read since its contents are, of necessity, interdisciplinary; there are elements of history, philosophy, physics, theology, and biblical studies. The material aims to bring to light our presuppositions and so provide a framework and a set of adaptable tools to address the concerns of congregants and skeptics. What is important, in my view, is a willingness to engage the material, and to have our Christian worldview challenged without equating that to undermining our core faith.

This is *not* another book that attempts to defend God from perceived threats—as if God needs *me* to defend him! Neither is it another response to the New Atheists. Nor is it a book that tries to show that the findings of modern science are in concord with Scripture. Rather, it is an attempt to learn wisdom from history—both ancient and modern—and to see how we can move beyond the old battlegrounds of modernity in what is now a postmodern world. The aim of this book therefore is to enhance faith. But in order to do that we have to first look at *foundations*; only then are we in a position to consider *issues*.

Too often conversations on science and faith skate over much deeper assumptions (or perceptions) on the nature and interpretation of Scripture, and of science. Instead, the rhetoric quickly goes toward issues, like evolution, global warming, or genetic engineering, without establishing a framework of mutual understanding in order for the dialogue to take place and be respectfully heard. The difference is not simply about the issues that are perceived to be at stake, but about the whole process on how such matters are

6. See, for example: BioLogos (biologos.org); the American Scientific Affiliation and their international partners (network.asa3.org); Test of Faith Project (www.testoffaith.com—from the Faraday Institute of Science and Religion); the Vatican Observatory (www.vaticanobservatory.va); and the Center for Theology and the Natural Sciences (www.ctns.org).

to be addressed. For example, what is the role of Scripture, reason, experience, and tradition in formulating our theological perspective? The balance of those four elements will differ depending upon your Christian tradition. It is naive to proclaim *sola scriptura* as if reason, experience, and tradition had no role to play at all. The Bible needs to be interpreted and that requires the use of reason. That process does not occur in isolation; rather our intellectual faculties have been shaped by our educational and religious traditions. Moreover our context in reading the Bible is very different from that of, say, the Middle Ages, and therefore our cultural experience also shapes the lenses through which we read Scripture. Add to that the whole topic of hermeneutics, which is the theory or principles of biblical interpretation, and we begin to see that Luther's *sola scriptura* is not as simple as it might seem at first glance. Indeed, if we look back at church history, we will see that Luther's conflict with Zwingli at Marburg (1529) over the nature of the Eucharist (Holy Communion) was—in part—because *sola scriptura* was not a sufficient enough criterion to resolve differences in interpretation.[7]

The first chapter revisits the battle over biblical interpretation at the time of Galileo. This is timely, since 2016 marks the four hundredth anniversary of the church's condemnation of heliocentrism, and there are still important lessons we must remember and relearn today. The key issue in 1616 was not so much about Galileo, but on *who* interprets Scripture and on *what* basis or principles. Chapter 2 squarely faces the issue of the inspiration and interpretation of Scripture. *Who* says the Bible is inspired? What are the *principles* for interpreting Scripture in our postmodern context? And what is the *purpose* of Scripture anyway? Chapter 3 critically explores the nature of science, exposing the inherited stereotypes that we often perceive to be true, which have created and fueled the historical tension between science and religion. Having established foundations—or at least communicated my own assumptions—we are then in a position to consider ways of *relating* science and theology. Following Ian Barbour, four approaches will be outlined and discussed in chapter 4. Are science and Christianity essentially in *conflict*, or are they compartmentalized and hence *independent* from one another? What can we learn if they respectfully *dialogue* with each other? What can we discover about God by studying nature itself? Can science and Christianity eventually be harmonized or *integrated*, resulting in enlightened, ethically responsible science and new understandings and formulations of Christian doctrine? I come to this subject as a physicist,

---

7. See, for example, Lindberg, *European Reformations*, 181–87.

Introduction

so my questions are from that perspective. Those from other scientific disciplines, such as biology, genetics, geology, psychology, etc., have their own subject-specific concerns and challenges. Chapter 5 explores what I regard as a key issue, that of chance and order. Quantum mechanics, which describes atoms, molecules, and their constituents, radically challenges our commonsense view of a cause-and-effect world. This has resulted in a statistical description of nature at the microscopic level, shattering the previous mechanistic, clockwork view of the cosmos. What, as Christians, are we to make of the element of chance (indeterminacy) that seems to be at the heart of nature? Does even talking about the role of chance in nature fill us with fear because it challenges our desire for "control," if not by us, at least by an all-powerful God? Or does it fill us with excitement over a world that is pregnant with new possibilities? This raises questions as to the nature of God. What do we mean when we say God is omnipotent? This and other divine attributes will be explored briefly in chapter 6, with particular focus on God's relationship with *time*. In addition to our views on Scripture, what we assume about the nature of God shapes the conversation between science and faith. Chapter 7 shifts gears from the God who is the *Creator* and *Sustainer* of the cosmos to the *personal, loving* God whom Christians worship. I explore the plausibility of the miraculous and the notion of petitionary prayer. Finally, in chapter 8, I return to examining the biblical creation narratives in the light of this journey of discovery. In addition to briefly exploring the classic texts of early Genesis, other Old Testament creation texts are examined which present a complementary view of God's ongoing creative activity and emphasize God as Sustainer.

With a flexible framework and suitable tools that church leaders apply self-critically, we are in a positon to address a wide range of important, topical science-faith (and other) issues that go well beyond the scope of this book. If I make some headway in that direction, this project will have been a success. If some of the traditional myths (and even fears) associated with thinking about the Bible, or science, or both, have been dispelled (and fears eased), this book will have been a success. If Christians, jaded by the institutional church's viewpoints in this arena, find grounds for hope, then this book will have been a success. If your faith in God as Creator and Sustainer is enhanced, and if you are newly inspired to pray, this book will have been a success.

As you journey onward, reflect on the words of Lesslie Newbigin:

Introduction

Both faith and doubt are necessary elements in this adventure [of knowing]. One does not learn anything except by believing something, and—conversely—if one doubts everything one learns nothing. On the other hand, believing everything uncritically is a road to disaster. The faculty of doubt is essential. But . . . doubt always rests on faith and not vice versa.[8]

---

8. Newbigin, *Proper Confidence*, 24–25.

Chapter 1

Science and Scripture
The Bible on Trial

But I do not feel obliged to believe that the same God who has endowed us with senses, reason, and intellect has intended to forgo their use and by some other means to give us knowledge which we can attain by them. He would not require us to deny sense and reason in physical matters which are set before our eyes and minds by direct experience or necessary demonstrations. This must be especially true in those sciences of which but the faintest trace . . . is to be found in the Bible. —Galileo Galilei, *Letter to the Grand Duchess Christina* (1615)

You shall love the Lord your God with all your heart, and with all your soul, and with *all your mind*, and with all your strength. —Mark 12:30[1]

INTRODUCTION

February 1616, four hundred years ago, was a momentous month in the history of science and Christianity. It was then that Copernicus's heliocentric view of the universe was condemned by the Roman Catholic Church as being a heretical teaching and in contradiction to Holy Scripture. Much

---

1. Emphasis mine. See also Matt 22:37 and Luke 10:27.

has happened since then, of course, but it is both an appropriate and timely starting point, not least as many still hold the view that the Bible is in conflict with the findings of science.

The Trial of Galileo, as it is sometimes called, occurred in two distinct phases. In 1616, the heliocentric system itself was on trial, *not* Galileo. The second phase, in 1633, was far more personal. It sought to ascertain if Galileo had abided to the details and spirit of the 1616 injunction, and—if need be—to discipline him. Much has been written about "the Galileo affair," and here is not the place to review it, suffice to say that this was about power, politics, patronage, popes, precedents, principles, polemics, and personalities. The issues were complex and nuanced—and need to be seen in their historical context. For instance, one wonders if the issue would have erupted as it did if a certain Martin Luther had not publicized his ninety-five theses a century beforehand. That challenge to the Roman Catholic Church's authority resulted in the Counter-Reformation and the Council of Trent (1545–63)—highly relevant in defining the theological milieu of Galileo's day, but more on that later. What I want to focus on here is the place of the Bible in the 1616 ruling, and how it was interpreted. To set the scene, it is worth remembering a few major developments in the previous centuries.

Improved farming practices, including the earlier development of the heavy plow, enabled huge increases in food productivity in Northern Europe. This, and other factors, helped support increasing urbanization and so transformed the social landscape. With cities came centers of learning, and by 1200 CE the great universities of Western Europe, such as Paris, Oxford, and Bologna, were founded by the church authorities and local rulers. Logic, mathematics, astronomy, natural philosophy, music, art, and law were all part of the academic syllabus, as well as—of course—theology.

During the eleventh and twelfth centuries there were many works of the Greek philosophers that were translated from Arabic into Latin in northern Spain and southern France. These ancient texts, along with commentaries on them, were preserved and enhanced by Islamic scholars. It was like the discovery of a long-lost treasure, and these works transformed the medieval period. Their contents were devoured by the new universities.[2] Natural philosophy was greatly enhanced by these texts from antiquity,

---

2. You can also imagine that when Greek texts became available, many of which had been preserved in the Greek-speaking Byzantine Empire, they could be compared with these translations. Thus long-lost texts and their translations fueled the quest for a revival of Greek culture and birthed Renaissance Humanism.

especially the works of Aristotle. By the end of the fifteenth century, the Renaissance was well under way.

During this period there was another significant invention: the printing press. This made the distribution of new information and ideas much more rapid. One of the first books printed was the Gutenberg Bible between 1450–56. Furthermore, in 1492 Columbus discovered the New World and in 1497, Vasco da Gamma rounded the Cape of Good Hope and so opened a trade route to India and the Far East. A new era of exploration and discovery had begun. These events fired the imagination of Europeans, and their mental horizons of time and space began to expand again. Dramatic developments emerged in all areas of human activity such as religion, art, music, and science.[3]

Having identified a few pertinent, historical highlights, and so set the scene, it is now necessary to outline the general way the Bible was interpreted at the time. This will be considered briefly in the next section, along with the then relationship between science—or natural philosophy—and theology.

NATURAL PHILOSOPHY AND BIBLICAL INTERPRETATION

How was one to understand Aristotle's view of the world with that given in Genesis? This requires the Bible to be interpreted and it is naive to think that biblical interpretation is somehow self-evident. McGrath states: "There is a sense in which the history of Christian theology can be regarded as the history of biblical interpretation."[4] An essential part of that long history is, therefore, addressing the question: "What were the accepted principles for biblical hermeneutics and exegesis, and how have they evolved?"[5]

The foundations of biblical interpretation began in the patristic period with different schools of thought emerging from the various Christian centers, such as Alexandria and Antioch. In addition to a literal interpretation, there emerged significant emphasis on allegorical interpretations, or

---

3. For example, major new styles in art and sculpture were being developed by men such as Michelangelo (1475–1564), Leonardo da Vinci (1452–1519) and Raphael (1483–1520), all contemporaries of Martin Luther (1483–1547).

4. McGrath, *Science and Religion*, 3.

5. "Hermeneutics" is the general theory, principles, or methodology of interpretation, and "exegesis" is the critical explanation or interpretation of a specific text.

hidden spiritual meanings, such as the Song of Songs corresponding to the love between Christ and his church. By the Middle Ages there was a standard method of biblical interpretation with four elements, namely: a *literal* sense of Scripture, in which the text was simply taken at face value, and three *nonliteral* approaches: *allegorical* (a mystical or metaphorical sense), *tropological* (a moral sense), and *anagogical* (a future sense). As we will see later, this elaborate characterization is an important element in the 1616 decree.

Returning to the patristics, given the importance of St. Augustine (354–430) it is worth quoting from his commentary on Genesis on the relationship between science, faith, and the Bible:

> In matters that are so obscure and far beyond our vision, we find in Holy Scripture passages which can be interpreted in very different ways without prejudice to the faith we have received. In such cases, we should not rush in headlong and so firmly take our stand on one side that, if further progress in the search for truth justly undermines our position, we too fall with it. We should not battle for our own interpretation but for the teaching of the Holy Scripture. We should not wish to conform the meaning of Holy Scripture to our interpretation, but our interpretation to the meaning of Holy Scripture.[6]

Consequently, whatever can be reasonably established through natural philosophy—and, of course, Augustine was very aware of the works of Plato, Aristotle, and other Greek philosophers—should not be an unnecessary source of contention in biblical interpretation or undermine or jeopardize the faith. Augustine therefore advocated the avoidance of intransigence in biblical interpretation on matters not central to the faith. Later, the influential Thomas Aquinas (1225–1274) cites Augustine as teaching:

> The first is, to hold the truth of Scripture without wavering. The second is that since Holy Scripture can be explained in a multiplicity of senses, one should adhere to a particular explanation, only in such measure as to be ready to abandon it, if it be proved with certainty to be false; lest Holy Scripture be exposed to the ridicule of unbelievers, and obstacles be placed to their believing.[7]

---

6. Cited in McGrath, *Science and Religion*, 5–6.

7. Aquinas, *Summa Theologica*, pt. 1, q. 68, art. 1, 762. This is in the context of God's creative work on the second day.

Science and Scripture

The concern was evidently to defend Scripture from unnecessary derision as a consequence of reasonable, intelligent arguments. In the spirit of Ecclesiastes 3, there is a time to defend a particular scriptural interpretation, and a time to let go if rational evidence demands it—else faith itself is not served.

Many of the patristic fathers developed the view that science and philosophy were "handmaidens to theology," building on the earlier idea of the Jewish scholar Philo of Alexandria. This handmaiden approach was also adopted by St. Augustine, meaning that Greek philosophy could be of use in *serving* theology—a view that is implicitly held by Christian apologists today. As mentioned earlier, when Aristotle's natural philosophy came to Europe via the translated texts, it was embraced by university academics.[8] However, Aristotle's views of the world clashed in places with those of conservative theologians within the church, and became a controversial topic—especially at the University of Paris. Ultimately some of Aristotle's works were banned in 1277. It is in this context that the great figure of Thomas Aquinas made his enormous contribution to Scholastic theology.

According to Edward Grant, over time—and in no small way due to the brilliance of Aquinas (and others)—natural philosophy and theology became independent disciplines.[9] This allowed science to be studied for its own sake, no longer tied in the service of theology as its handmaiden. Theology's superior status as "queen of the sciences" was able to pacify the church and many theologians were still able to make use of natural philosophy in their understanding of the Christian faith. Grant concludes that "while natural philosophy was virtually independent of theology, theology was utterly dependent upon natural philosophy."[10] Consequently, by the late Middle Ages, Aristotelianism became absorbed within the Christian worldview. This perspective was still dominant among theologians and within academia at the time of Galileo. Galileo was therefore combating the physics of Aristotle, which Galileo had shown to be in error by means of experiment, but which was now embodied within the church's theology. Thus Aristotle's natural philosophy, elements of which had been theologically

---

8. Edward Grant discusses at length the relationship between natural philosophy and theology in the late Middle Ages; see Grant, *History of Natural Philosophy*, 239–73.

9. Grant, *Science and Religion*, 184–90.

10. Grant, *History of Natural Philosophy*, 273.

controversial at the time of Thomas Aquinas, had become part of the conservative establishment.[11]

THE LANGUAGE OF ACCOMMODATION

Returning to the Galileo era, we can appreciate that in a *literal* reading of Genesis 1, God's six days of creative activity would have been understood as six periods of 24 hours. A *nonliteral* approach could see the opening chapter of Genesis as poetic allegory. A third approach is that of *accommodation* in which revelation is divine condescension to a culturally appropriate level, with its language and norms, such that it could be readily understood by the original audience. This latter approach has a long history and is particularly important in considering the relationship between science and theology today.

The language of accommodation was expressed as early as Origen (ca. 185–254): "God condescends and comes down to us, accommodating to our weaknesses, like a schoolmaster talking a 'little language' to his children, or like a father caring for his own children and adopting their ways."[12] St. Bonaventure (1221–1274), a contemporary of Aquinas, declared the same sentiments: "Scripture, condescending to poor, simple people, frequently speaks in a common way."[13] Accommodation, however, is often associated with the Protestant reformer John Calvin (1509–1564). When studying Scripture, Calvin (like Luther) saw the biblical text primarily through a "Christological lens" and broadly rejected the allegorical interpretations of the Roman Catholic Church, opting instead for the plain or straightforward meaning of the text.[14] Naturally, for that time period, he accepted the words from Scripture to be true.[15] However, Calvin developed the earlier

---

11. David Lindberg describes Aquinas as Christianizing Aristotelianism and at the same time "Aristotelianizing" Christianity. See Lindberg, *Beginnings of Western Science*, 233–34.

12. Cited in McGrath, *Christian Theology*, 192.

13. Cited in Grant, *History of Natural Philosophy*, 269. Elsewhere Grant states that Nicole Oresme (1320–1382) adopted a similar attitude; see Grant, *Science and Religion*, 223.

14. Dillenberger, *Protestant Thought and Natural Science*, 30–31; Gerrish, "Reformation and the Rise of Modern Science," 256–57.

15. The modern notion of the "infallibility of Scripture" was not a concept that he (or Luther) considered. Moreover, Calvin did not defend the old astronomy in light of a "developed conception of the inerrancy of Scripture." See Dillenberger, *Protestant Thought*

understanding of God's *accommodation* or condescension to humanity in describing God's inspiration of the biblical authors.[16] Calvin writes (concerning Ps 136:7):

> *The Holy Spirit had no intention to teach astronomy*; and, in proposing instruction meant to be common to the simplest and most uneducated persons, *he made use by Moses and the other Prophets of popular language, that none might shelter himself under the pretext of obscurity,* . . . *the Holy Spirit would rather speak childishly than unintelligibly to the humble and unlearned.*[17]

The advantage of accommodation is that it provides a cushion against strict literalism in biblical interpretation. The creation texts, for example, may well be describing physical reality, but not in a technical sense or in the way God himself regards the cosmos. Rather, the language used (by the Holy Spirit) is greatly simplified so that the general public can comprehend the essential message.

In Calvin's *Institutes*, arguably his most-considered work given its number of revisions, he writes:

> Therefore, in reading profane authors, the admirable light of truth displayed in them should remind us, that the human mind,

---

and Natural Science, 30–31, 38.

16. The root of Calvin's accommodation is debated; it could be via Augustine/Chrysostom/Erasmus, or more generally through his humanist education; see Balserak, "Exegesis and *Doctrina*," 377.

17. Calvin, *Commentary on the Psalms*, 5:168, emphasis mine. The same theme of accommodation occurs in Calvin's commentary on Ps 19: "[David] shows us the sun as placed in the highest rank, because in his wonderful brightness the majesty of God displays itself more magnificently than in all the rest. *The other planets, it is true, have also their motions, and as it were the appointed places within which they run their race, and the firmament, by its own revolution, draws with it all the fixed stars, but it would have been lost time for David to have attempted to teach the secrets of astronomy to the rude and unlearned; and therefore he reckoned it sufficient to speak in a homely style*, that he might reprove the whole world of ingratitude, if, in beholding the sun, they are not taught the fear and the knowledge of God. . . . *He does not here discourse scientifically (as he might have done, had he spoken among philosophers) concerning the entire revolution which the sun performs, but, accommodating himself to the rudest and dullest, he confines himself to the ordinary appearances presented to the eye*, and, for this reason, he does not speak of the other half of the sun's course, which does not appear in our hemisphere" (Calvin, *Commentary on the Psalms*, 1:314, emphasis mine). Note how Calvin differentiates between how God might have spoken to a scientist or philosopher on the heavens, and to ordinary mortals. Moreover, God makes no attempt to teach David the secrets of heavenly motion, simply to view them as a pointer to the Creator.

however much fallen and perverted from its original integrity, is still adorned and invested with admirable gifts from its Creator. *If we reflect that the Spirit of God is the only fountain of truth, we will be careful, as we would avoid offering insult to him, not to reject or condemn truth wherever it appears.* In despising the gifts, we insult the Giver.[18]

In using the words "profane authors," Calvin is referring to the "pagan" philosophers from antiquity! Nevertheless, here we have a positive view of astronomers and astronomy as enhancers of our glimpse of God's glory. The nobility of the mind, regardless of the Fall, leads Calvin to affirm Augustine's truism: "Let every good and true Christian understand that wherever truth may be found, it belongs to his Master."[19] This quote is often paraphrased as: "All truth is part of God's truth."

SETTING THE STAGE: THE COUNCIL OF TRENT (1545-1563)

Following Martin Luther's break with Rome in 1519, it was necessary for the Roman Catholic Church to reassert its authority. This it did with the Council of Trent which, among other things, addressed the status of Scripture, revelation, and tradition. One decree concerning the principle of church tradition states:

> The Council also clearly maintains that these truths and rules are contained in the written books *and in the unwritten traditions* which, received by the apostles from the mouth of Christ himself or from the apostles themselves, the Holy Spirit dictating, have come down to us, transmitted as it were from hand to hand. Following then the examples of the Orthodox fathers, it receives and venerates with a feeling of *equal piety* and reverence *both* all the books of the Old and New Testaments, since one God is the author of both, *and also the traditions* themselves, whether they relate to *faith or to morals*, as having been dictated either orally by Christ or by the Holy Spirit, and preserved in the Catholic Church in unbroken succession.[20]

---

18. Calvin, *Institutes*, bk. 2, ch. 2, 15, 170.
19. Augustine, *On Christian Doctrine*, II, 18, 28.
20. Blackwell, *Galileo, Bellarmine, and the Bible*, 9, emphasis mine.

This statement gives a sense of how the Bible and the "unwritten traditions" (i.e., traditions outside of Scripture) were viewed. The views of the venerated fathers of the church were regarded with "equal piety and reverence" as they too were deemed to be inspired by the Holy Spirit by the Roman Catholic Church. This Counter-Reformation rhetoric was intended to respond to Luther's views. Instead of *sola scriptura*, we see the dual emphasis on Scripture *and* tradition. The council was emphatically reestablishing the authority of the "true" church in all matters relating to faith or morals.[21] Another decree states:

> Furthermore, to control petulant spirits, the Council decrees that, in matters of faith and morals pertaining to the edification of Christian doctrine, *no one, relying on his own judgment and distorting the Sacred Scriptures according to his own conceptions, shall dare to interpret them contrary to that sense which Holy Mother Church*, to whom it belongs to judge of their true sense and meaning, has held and does hold, or even contrary to the unanimous agreement of the Fathers.[22]

This declaration is also to be understood in the context of the Reformation with its emphasis on individual biblical interpretation. Such "petulant spirits" could not be tolerated! This statement would also certainly be relevant to Galileo personally, and to the heliocentric worldview. Only the Roman Catholic Church can judge the true sense and meaning of Scripture. The pronouncement, however, does not specify on what basis—or principles—the Bible can be interpreted, but it will certainly involve being consistent with the teachings of the traditional church fathers. This latter aspect adds further complexity because it is not just scriptural interpretation of a given text that was involved, but the legacy of church tradition. Even if it could be argued that a text should be taken in a nonliteral way, could it be demonstrated that the church fathers—from the patristics onwards—*also* viewed it as nonliteral? This creates tremendous exegetical inertia thereby inhibiting change.

An influential commentary on Genesis by Benito Pereyra (1535–1610) emphasized that biblical interpretation should be taken literally and historically. Blackwell comments that "this increasing emphasis on literalism

---

21. Blackwell points out that "morals" is a more general term which also includes the contents of the canon, the edition and translations of Scripture, the legitimacy of councils and elections, and the determination of the sacrament of ordination. Ibid., 12–13.

22. Ibid., 11–12, emphasis mine.

was characteristic of the Counter-Reformation response to Trent."[23] However Pereyra also agreed with Augustine and advised avoiding conflict with anything that can be firmly established from natural philosophy: "The truth of sacred Scriptures cannot be contrary to the true arguments and evidence of the human sciences." Galileo was clearly aware of this commentary and used its material in writing his famous *Letter to the Grand Duchess Christina* in 1615.[24]

CARDINAL ROBERTO BELLARMINE AND THE BIBLE

Cardinal Bellarmine (1542–1621), a formidable theologian of his day, was the key person behind the 1616 decision to condemn heliocentrism. He was intimately involved with the Counter-Reformation "wars" and actively worked on the revised version of Jerome's Latin Vulgate, known as the Clementine edition. In Bellarmine's *Controversies* there is a section on "The Word of God." Among other things he stated: "Scripture is the immediately revealed word of God and was written as dictated by God."[25] Consequently Scripture is deemed "inerrant," to use a more modern term. He continues:

> In Scripture there are many things which of themselves do not attain to the faith, that is, which were not written because it was necessary [for salvation] to believe them. But *it is necessary to believe them because they were written.*[26]

That last clause is telling. This goes beyond the "faith and morals" in the Council of Trent's statement; instead, it is necessary to believe something simply because it is there in the Bible. This logically follows from the biblical authors being directly inspired by God—and God never lies or deceives.

---

23. Ibid., 21.
24. Ibid., 22.
25. Ibid., 31.
26. Ibid., 32, emphasis mine. An alternate translation of the last clause is: "but those things are necessarily believed which are written." See Bellarmine, *Disputations*, 431. The context is as the title suggests: the unwritten tradition is just as much the word of God as is Scripture; ibid., 344, 432. Furthermore, Bellarmine states: "The total rule of faith is the word of God, or the revelation of God made to the Church, which is divided into two partial rules, Scripture and Tradition. And indeed, Scripture, because it is a rule, has from this that *whatever it contains is necessarily true and to be believed*" (ibid., 432, emphasis mine). Bellarmine's logic is this: Whatever God has revealed in Scripture is true; God has revealed this in Scripture, therefore this is true (ibid., 336).

Since this was the view of Galileo's opponent, there was inevitably going to be a clash of perspectives.

The Bible and how it was to be interpreted was, therefore, at stage center in 1616. Bellarmine recognized both literal and nonliteral interpretations of Scripture, the latter with the three subcategories mentioned earlier. One would therefore think there could be room for debate over whether early Genesis should be understood literally or figuratively. However there was another famous theological duel, this time between two Protestants at Marburg in 1529. It was between Huldrych Zwingli and Martin Luther over the Christ's words in Matthew 26:26 "this is my body" and the nature of the bread in Holy Communion. Luther favored a literal interpretation where Zwingli favored a figurative one.[27] Bellarmine would have totally agreed that this phrase should be understood literally. He therefore recognized the exegetical dangers of trying to introduce figurative interpretations when the literal one was already well-established in church tradition.[28]

This provides the context into how one should understand the heliocentric-geocentric duel on which Bellarmine adjudicated.[29] The Bible speaks of the earth being at rest (e.g., Ps 93:1; 96:10; 104:5; 1 Chr 16:30; Eccl 1:5); should this be understood literally or figuratively? How were these texts universally understood by the church fathers? Furthermore, there was another major issue: *who* had the right to determine the interpretation of Scripture? It certainly would not be Galileo, but the Roman Catholic Church. As Blackwell summarizes:

> The individual judge of Scripture faced a double jeopardy; one relating to the *content* of the interpretation, the other to assuming the *role* of being an interpreter. No matter what the merits of the former, the individual was always in jeopardy on the latter.[30]

---

27. There is much more nuance to this incident concerning the *meaning* of words. For example, both sides rejected the Roman Catholic doctrine of transubstantiation, yet both agreed with *sola scriptura*! For further details, see Lindberg, *European Reformations*, 181–87.

28. Blackwell, *Galileo, Bellarmine, and the Bible*, 35.

29. In the 3rd edition of Langford's book, he comments: "Blackwell's treatment of the differences between Galileo and Bellarmine on Biblical interpretation and natural scientific knowledge . . . are incisive and exactly right." Langford, *Galileo, Science and the Church*, 192.

30. Blackwell, *Galileo, Bellarmine, and the Bible*, 36–37.

SCIENCE AND CHRISTIANITY

GALILEO, FOSCARINI, AND THE DECISION OF 1616

Copernicus's book was published in 1543 and for some seventy years there was little reaction within the Roman Catholic (and Protestant) Church. Yet within a period of about two years, around four hundred years ago, Copernicus's heliocentric worldview became condemned. Given the slow pace at which any institutional body normally moves, what happened to create such a rapid response in the church?

In part, of course, it was due to the great advances in astronomy with the earlier work of Tycho Brahe (1546–1601), Johannes Kepler (1571–1630), and Galileo's experimental observations using the then recently invented telescope. By the end of 1610, he had observed the four largest moons of Jupiter, demonstrating that there was more than one center of rotation in the universe.[31] He had also seen the phases of Venus, which demonstrated that that planet (at least) must be moving around the sun—rather than around the earth. In 1613 Galileo had published his work on sunspots and shown that the sun itself was rotating. In addition to the sun's blemishes, the earlier supernova of 1604 also created debate since according to the wisdom of the day, everything beyond the moon was deemed to be perfect and unchangeable. That Aristotelian worldview of the heavens was being challenged within university circles by Galileo's experimental observations, all of which were confirmed by Jesuit astronomers. However, since this worldview had been absorbed within the Christian theology, it was bound to impact ecclesiastical circles as well.

A problem arose when Foscarini, a Carmelite priest and a trained theologian, published a short book in 1615 aiming to demonstrate that heliocentrism was *not inconsistent* with the Bible. Moreover, he emphasized that biblical interpretation on the key texts should not be too definitive because if, in the future, heliocentrism is shown to be true, then it could give the impression that the Bible contains errors.[32] In addition to also endorsing the principle of the accommodation, Foscarini made another key point:

> The Church . . . cannot err in matters of faith and salvation only. But the Church can err in practical judgments, in philosophical

---

31. One objection to a moving earth, based on an Aristotelean understanding of motion, was that it would leave the moon "behind." Galileo's observation of the moons of Jupiter demonstrates that a planet can move without losing its moons!

32. Ibid., 89.

speculations, and in other doctrines which do not involve or pertain to salvation.[33]

Foscarini sent a copy of his letter to Cardinal Bellarmine for his evaluation and whose studied response was informal but authoritative. Bellarmine said there is "nothing dangerous" treating heliocentrism as a mathematical model, but it would be a "very dangerous thing" to treat it as a real, physical description of the world, as it would be destructive to faith by making out the Scriptures to be false.[34] Bellarmine also reasserted that the mere fact that something is present in the Scriptures makes it not only certainly true, but also a matter of faith—assuming that its meaning is clearly established. Whatever the Council of Trent had decreed, Bellarmine was using his own principles of interpretation in determining how to understand matters of "faith and morals."

Reaction to Galileo's *Letters on the Sunspots*, where he openly declares his support for Copernicus, was not long in coming.[35] Galileo's friend and collaborator Benedetto Castelli, a Benedictine monk and professor of mathematics at Pisa, informed Galileo that he had been party to a dinner conversation at the Tuscany court. During the conversation, another distinguished professor of philosophy at Pisa (Cosimo Boscaglia) had pointed out that Galileo's discovery of the existence of four moons orbiting around Jupiter was indeed true, but that nevertheless the earth remained motionless because to be otherwise would be contrary to the Bible. Castelli had defended Galileo's position but despite the discussion, de Medici's

---

33. Ibid., 234–35.

34. Ibid., 104. Things are even more complicated! Bellarmine was aware of the (unsigned) preface in Copernicus's *De Revolutionibus* and pointed out that this demonstrated that Copernicus himself did not *really* believe that the earth moved, but that his system was simply a helpful mathematical model. If that is what Copernicus believed, what new incontrovertible evidence did Galileo have to prove otherwise? There was no such evidence and Tycho Brahe's system fitted all the known facts. Consequently, there was no problem in the Roman Catholic Church accepting the Copernican system as a working *hypothesis*, but not as fact. The infamous unsigned preface was, in fact, written by Lutheran theologian Andreas Osiander, who published *De Revolutionibus* in Protestant Nuremberg. Maestlin—Kepler's mentor—and no doubt other scholars and close friends of Copernicus, were aware that the anonymous preface was not written by Copernicus himself. Kepler also knew, and seventy years later made the truth public in his *Astronomia Nova* (1609). See Gingerich, *Book Nobody Read*, 141–42, 159–60.

35. This book was written in Italian, not Latin, so ensuring a wide reading audience. Galileo also received a letter of congratulations from Cardinal Barberini (later to become Pope Urban VIII).

mother—the Dowager Grand Duchess Christina—remained concerned. Galileo, not present at the dinner, felt it necessary to explain his position on the matter in detail. It began in the form of a *Letter to Castelli* and, a year later, the extended and now famous *Letter to the Grand Duchess Christina* (1615) addressed to his patron's family.[36] Whether these letters were private or not, they received wide circulation. In doing so, Galileo moved toward biblical exegesis—and so the famous controversy begins.

In these two *Letters*, Galileo pointed out that since God is both the author of nature and Scripture, the truths from both areas cannot contradict each other.[37] Galileo agreed that the Bible was inerrant, but then added that later interpreters can and do make mistakes. Galileo reminded his reader that God used the language of accommodation in Scripture. He also used the ideas of Augustine and Aquinas that, on matters concerning the natural order, scriptural interpretation should defer to the findings of science when they are firmly established from experimental and rational proof. The *intention* of Scripture was to give the knowledge needed for faith and salvation. As Cardinal Baronius (1538–1607) put it: "The Holy Spirit intended to teach us how to go to heaven, not how the heavens go." Moreover, Baronius was echoing the same intent of the Holy Spirit as Augustine.[38] This last argument seems to suggest that there should be a clear distinction—if not a total separation—between theology and natural philosophy.[39] The purpose of Scripture is God's revelation of salvation, not science, and consequently Galileo did not see it as a theological problem to adhere to the Copernican picture of the universe. However, in other parts of Galileo's writings, he—like Foscarini—tried to show that Copernicanism was *consistent* (or not inconsistent) with Scripture. *Concordism* is not the same as *separatism*, since the former inevitably involves biblical interpretation and for the latter interpretation is irrelevant. This distinction will be considered in detail in chapter 4.

Another problematic aspect was that, despite Galileo's astronomical observations, the heliocentric model was not able to be demonstrated as scientific "truth" at that time. All the experimental evidence was also consistent with Tycho Brahe's model of the cosmos. In that model all the

---

36. While the *Letter to the Grand Duchess Christina* was widely circulated, it was not published until 1635 in Strasbourg, after Galileo's trial in 1633.
37. Blackwell, *Galileo, Bellarmine, and the Bible*, 66.
38. Langford, *Galileo, Science and the Church*, 65.
39. Barbour, *Religion and Science*, 14.

planets orbited the sun, but the sun—"carrying" all the planets—then orbited a stationary earth. We today might think this model somewhat contrived and trying to preserve the geocentric perspective. Nevertheless, this model, along with that of Ptolemy, needed to be demonstrated scientifically as false. This Galileo could not do.[40] Yet he was convinced that he was close to proving it and that it would be possible for heliocentrism to be proved *conclusively* at some point in the future.[41] Many theologians, however, thought that God's universe was too complicated and mysterious for us to *ever* have *certain* knowledge of such matters. Consequently, in the absence of what we commonly today term "absolute proof," the authority of the church, Scripture, and Christian tradition would remain.

The conclusion of the 1616 "trial" declared unequivocally that the heliocentric worldview was *heretical* because it explicitly contradicts sacred Scripture (deemed rightly understood), the understanding of church tradition, and learned theologians. It can be argued that Galileo wanted the church to *return* to the tradition of Augustine and Aquinas concerning the relationship between natural philosophy and Scriptural exegesis. But in the context of the Counter-Reformation, it was impossible for church authorities to tolerate Galileo telling them how to interpret Scripture. However, if a person was to consider the Copernican system as a hypothesis, a mere mathematical model—but not a physical model—then one could express one's views without being in conflict with the church. The later 1633 trial of Galileo was, in part, on whether Galileo could adhere to that perspective.

As we have seen, the nuanced issues were far more complicated than those who would simply portray the *ultimate* outcome of the trial of Galileo as the triumph of science and reason over faith and dogma. However, that does not excuse the church for this tragically wrong, yet pivotal, decision in the history of science. It does, however, demonstrate the tangle of faith, science, Scripture, and reason—a tangle that continues to this day. In the

---

40. If the earth orbited the sun, the relative movement of nearby stars should be discernable with respect to more distant stars, yet Brahe and Galileo could not measure this stellar parallax. Had that been technically possible at the time, then Galileo would have "demonstrated" that the earth moves. This matter was not fully resolved until Bessel observed stellar parallax in 1839, following James Bradley's earlier discovery of the "aberration of light" in 1728.

41. Galileo thought (or claimed) he had established incontrovertible proof for the Copernican worldview through his now infamous flawed reasoning for the existence of the tides, which Kepler correctly realized were due to the moon.

next section I want to focus on what lessons have been—and still can be—learned by the church today.

GALILEO: LESSONS FOR TODAY

George Santayana famously said: "Those who cannot remember the past are condemned to repeat it."[42] There are times when I wonder if some Christians have fully grasped the significance of the Galileo "trial," because, in certain quarters at least, the science and Christianity debate has a strong sense of *déjà vu* about it. The background and issues in the 1616 edict have been related at length so that many of the subtleties could be appreciated afresh. We now begin to explore what has been learned and, more importantly, consider what we can still discover and appropriate today.

One obvious consequence is demonstrated by the development and the character of modern universities, which took a more secular route during the Enlightenment and largely rejected the establishment authorities. Although clerics populated and developed the early universities, things are very different today. Inevitably, science became totally independent from ecclesiastical authority.

But what about the church? Have lessons been learned since the time of Galileo? In many ways, yes. Slowly, the Roman Catholic Church has come to a point where it has embraced the findings of Galileo and Pope John Paul II exonerated him in 1992.[43] Nevertheless, there were other casualties along the way. For example, the Jesuits, who were scientific pioneers at the time of Galileo and confirmed many of his astronomical observations, were effectively curtailed in their scientific creativity after the trial for fear of falling foul of their vow of obedience to their order.[44] Nevertheless the Roman Catholic Church has become very progressive in this area and embraced evolution and other established findings of modern science. The relationship between science and faith is not at all what it once was in the Roman Catholic Church.

However things are very different in the conservative evangelical church. Although history never repeats itself exactly, there are contemporary

---

42. Which is remarkably similar to Edmund Burke's earlier quote: "Those who don't know history are doomed to repeat it."

43. Ibid., 15.

44. Blackwell discusses this at length; Blackwell, *Galileo, Bellarmine, and the Bible*, 135–64.

issues that parallel those at the time of Galileo, and these bear reflecting upon. For example just as the Roman Catholic Church was being challenged by the Reformers, resulting in a climate of biblical literalism during the Counter-Reformation, so there is a connection between biblical literalism (especially among Christian fundamentalists) and the climate of fearful suspicion toward modernism. The main tool to address that threat is to insist upon the literal truth of Scripture, just as in 1616, but interpreted by Evangelicalism in this case. As such the nature of Scripture and exegetical methodology are deemed unassailable, and consequently anything that challenges them is attacked. An example of this stance is embodied in the 1978 *Chicago Statement on Biblical Inerrancy*. There are aspects of these articles that strongly echo Bellarmine's view of Scripture, such as:

> We affirm that Scripture in its entirety is inerrant, being free from all falsehood, fraud, or deceit. *We deny that Biblical infallibility and inerrancy are limited to spiritual, religious, or redemptive themes, exclusive of assertions in the fields of history and science.* We further deny that scientific hypotheses about earth history may properly be used to overturn the teaching of Scripture on creation and the flood.[45]

This declaration can be compared to, for instance, the 1647 Westminster Confession:

> All things in Scripture are not alike plain in themselves, nor alike clear unto all; yet *those things which are necessary to be known, believed, and observed, for salvation,* are so clearly propounded and opened in some place of Scripture or other, that not only the learned, but the unlearned, in a due use of the ordinary means, may attain unto a sufficient understanding of them.[46]

Among other things, this statement has resonances with Cardinal Baronius's and Augustine's views on the Holy Spirit's intention for Scripture mentioned earlier, i.e., *sufficient for salvation*. This serves to highlight that, regardless of biblical interpretation or hermeneutics, *the very purpose or*

---

45. *Chicago Statement on Biblical Inerrancy*, art XII, emphasis mine. The Chicago Statement was signed by nearly three hundred noted evangelical scholars, including Norman L. Geisler, J. I. Packer, Francis Schaeffer, and R. C. Sproul. See also Summary Statement No. 4: "Being wholly and verbally God-given, Scripture is without error or fault in all its teaching, no less in what it states about God's acts in creation, about the events of world history, and about its own literary origins under God, than in its witness to God's saving grace in individual lives."

46. *Westminster Confession*, ch. 1, no. 7, emphasis mine.

domain of Scripture needs to be defined or articulated in order to have a meaningful dialogue. This foundational difference is but one example why Christians can be talking at cross purposes on the matter of science and faith. I will address this further in the next chapter.

The parallels between then and now continue; consider what happens when there is dissent. For example, a minister in a congregational church, or a professor in a Christian university, can be dismissed from their respective positions for a difference in theological opinion from the accepted norm. In other words, there are governing bodies, or boards of directors, who are willing—and able—to provide strict theological censure to those who are deemed to be "heretical." I am not arguing for no accountability or boundaries, rather sober reflection on the overuse of strong rhetoric of "heretic" (or its equivalent), a term often being used pejoratively to empower the authority. Differences of opinion are to be *expected* in the body of Christ (1 Cor 12), rather than avoided by all means possible.

Even though the Roman Catholic injunction against the heliocentric worldview suppressed further publishing on this matter, it did not inhibit people's thoughts, or further exploration on this topic. Lips may be sealed and books may be banned, but ideas, questions, and doubts cannot be suppressed in this way. Moreover, and with the benefit of hindsight and reflection, how does such an injunction truly honor God's reputation and enhance the Christian faith? Instead, those very two things were undermined by this heavy-handed approach, which was more about power, politics, and principles.

One can be sympathetic to the cause of respecting the authority of Scripture and that of the church, but we should not think that God himself needs to be defended, especially by us. Ultimately, the Roman Catholic Church lost its influence among the intelligentsia of Europe. Similarly, those today who attempt to establish their authority by exerting their power could well lose out in the end and do irreparable damage to their cause. I think the same has happened to Evangelicalism. In many quarters, Evangelicalism—or some branches of it—has lost its credibility. More to the point, it is persisting in fighting an old battle against modernism in a world that has, in many ways, already moved on to postmodern perspectives. In which case, this battle becomes irrelevant. Sadly the very book Evangelicals want everyone to respect becomes tarnished in the process as—to change the metaphor—undiscerning skeptics simply throw out the baby with the bathwater. Or, more precisely, the Bible is discarded because of a particular interpretation. As we saw earlier, this is exactly what Aquinas, citing Augustine, warned against.[47] In the face of centuries of scientific evidence, the Bible has become an object of doubt, disregard, or disdain—instead of being honored and trusted. Sadly, some Christian traditions are simply fighting an outdated war with the wrong tools.

Curiously enough, the Old Testament prophets were generally unpopular people because they challenged the establishment status quo. Some of those prophets were influential people closely associated with the monarchy or royal court; others were rural peasants or outsiders. Some were recognized in their own time as God's messengers and others were understood to be true prophets with the benefit of hindsight (which is one reason why their books were preserved). But if God used prophets as the conscience of the Jewish nation then silencing the voice of the prophet is a dangerous and counterproductive thing to attempt to do. If influential prophetic voices are not heard in the church today then the Holy Spirit may need to use "outsiders" to challenge the ecclesiastical status quo. It can be argued that Galileo, a devout Christian, was such a prophetic voice who was silenced. How much the church needs both courageous prophets to speak up on a wide range of contemporary issues and the discernment of Christian leaders to hear unpopular messages that challenge Christ's church today!

There were, in 1616, Jesuits and other secret Galileo sympathizers who recognized that the observational evidence had created some legitimate doubt on certain aspects of the Ptolemaic system. They may not have wanted to adopt a full-blown heliocentric universe, perhaps preferring instead Brahe's model, but they recognized that the telescopic observations were revealing genuinely new things that were challenging the Aristotelian worldview. In a similar way, there are moderate evangelicals and mainstream Christians who are able to embrace the findings of modern science without feeling that biblical Christianity itself is under threat.

Francis Collins, who also cites another of Augustine's exhortations for caution, writes:

> The scientific correctness of the heliocentric view ultimately won out, despite strong theological objections. . . . Could this same harmonious outcome be realized for the current conflict between faith and the theory of evolution? . . . Unfortunately, however, in many ways the controversy between evolution and faith is proving

---

47. As did Galileo's contemporary Foscarini, but he did not have the stature of Aquinas or Augustine.

much more difficult than an argument about whether the earth goes around the sun.[48]

He makes a fair point about the persistence of the evolution issue. However, I suggest that one of the issues that some Christians today have failed to grasp was a key underlying concern over the theological implications of a nonstationary earth at the time of Galileo. The difficulty with the heliocentric worldview, as opposed to geocentric, was that it *displaced humankind from the center of the universe*. Humankind was now drifting on one of the solar system's many planets orbiting around the sun. That problem is significantly worse today, as we now know that our sun is fairly nondescript and merely one of 100 billion stars in the Milky Way galaxy. And there are estimated to be 100 billion galaxies in the universe. A similar problem can be found within the evolution and creation debate. What is the status of humankind, made "in the image of God" (Gen 1:27), if we are an integral part of the animal kingdom? At the heart of the issues of heliocentrism and evolution is the implied question: *What, then, is the place of humankind in God's created order?* This anthropological question is one reason why the matter is so emotive.

Evidence that heliocentrism was profoundly destabilizing is given in the well-known lines of *An Anatomy of the World* (1611) by the English poet and cleric John Donne (1572–1631):

> And *new philosophy calls all in doubt*,
> The element of fire is quite put out,
> The sun is lost, and th'earth, and no man's wit
> Can well direct him where to look for it. . . .
> *'Tis all in pieces, all coherence gone*,
> All just supply, and all relation.[49]

Given the date of the poem, Stephen Toulmin—quite credibly—sees this excerpt as referencing Galileo's discoveries, along with those of Copernicus and Kepler.[50] The church's union of theology, geocentrism, and the Aristotelian worldview provided a robust framework whereby people knew their place in the cosmic structure. The implications of the "new philosophy" of heliocentrism to that worldview were keenly felt, as it cut humankind adrift from its traditional moorings: "The (Aristotelian) element of fire is quite

---

48. Collins, *Language of God*, 156–58.
49. Emphasis mine.
50. Toulmin, *Return to Cosmology*, 220.

put out." We should not underestimate the significance of those sentiments today; the depth of feeling—or of crisis—expressed in "'Tis all in pieces, all coherence gone" is self-evident. Once the earth was displaced so that it became one of the minor planets of the sun, instead of occupying the center of the cosmos, people lost their former sense of "knowing where they were" in the grand scheme of things.[51]

While everyone is quite at ease with the heliocentric worldview today—and most accept what astronomers say concerning our place in the vast cosmos—many conservative Christians still wrestle with the issue of evolution. One aspect of the matter is, I suspect, that we cannot really comprehend large numbers. We cannot *imagine* billions of stars and galaxies, or billions of years. These descriptors of both *space* and *time* are outside of our common experience. Nevertheless, many Christians are not uncomfortable with the vastness of space and so accept the big bang theory and the findings of modern astronomers. Yet, oddly, some are deeply troubled with the timescales of billions of years that is a feature of both cosmic and biological evolution. For me, the theological response to the above question on the spatio-temporal place of humankind in God's created order, in light of *both* cosmology and biological evolution, must be essentially the same. Differentiating between those two dimensions (and issues) is unnecessary, unhelpful, and unwise.

Before continuing it is worth reminding ourselves that the Scriptures do not argue for the geocentric perspective in the way that we commonly think. When we read Genesis 1:1, "In the beginning when God created the heaven and the earth," we imagine a NASA picture of a blue-green planet surrounded by space, stars, and the Milky Way galaxy. But that is not how the original audience would have imagined the created order. That verse could be translated: "In the beginning when God created the sky and the land"—and the rest of the chapter provides the details.[52] The biblical view of the created order was one that was essentially "flat" (or layered), with the underworld (*sheol*) below the land, and a domed canopy of sky above (see fig. 1). When we read certain biblical texts, like Isaiah 40:22, they become far more understandable when we imagine it from a *land*-based standpoint rather than our modern planetary perspective. When God created the world he did not even reveal it in Scripture to be ball-shaped; that perspective came later with the Greek philosophers. Even in the Greco-Roman world of the New Testament, we still

---

51. Ibid., 221.

52. For further discussion, see Winslow, "Earth Is Not a Planet," 13–27. See also Jonah 1:9; Ps 95:5.

get no hint that the earth is, in fact, spherical. It seems clear, then, that not even that minor detail was relevant in God's revelation to humankind. Although not mentioned in the Bible, the issue of whether the earth was round or flat was certainly being discussed by the early church fathers.[53] Some scholars, like Lactantius, rejected the Greek idea of a spherical planet. However, no one today seriously believes in the ancient Hebrew conception of the universe. Regardless of the *shape* of the earth, it was still deemed *central* and *stationary*. The place of humankind in God's created order was, therefore, not seriously threatened or challenged, only one's approach to biblical interpretation.

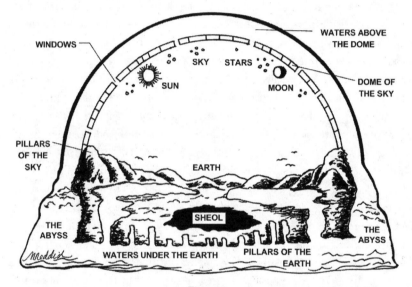

Figure 1.

How, then, should we approach interpreting the Bible today? That is the topic of the next chapter.

---

53. Grant, *God and Reason in the Middle Ages*, 341. Notice again how our *experience* of the world around us informs our physical worldview. A "flat" world is limited by the distant horizon. With no appreciation of the water cycle, the notion of waters above and below the earth is quite coherent with rain and floods. (Recall that the River Nile flooded annually.) With the introduction of the spherical planet by the Greek philosophers—and with no appreciation of gravity—the question as to whether people existed in antipodes was a moot point; after all, what is to stop them from "falling off"? In the same way, our *experience* of a stationary planet (and we still speak of sunrise and sunset as if the sun actually moves) was an intuitive argument against a heliocentrism; if the planet is moving, won't we be left behind?

Chapter 2

On the Inspiration and Interpretation of Scripture

> For as the rain and the snow come down from heaven, and do not return there until they have watered the earth, making it bring forth and sprout, giving seed to the sower and bread to the eater, so shall my word be that goes out from my mouth; it shall not return to me empty, but it shall accomplish that which I purpose, and succeed in the thing for which I sent it. —Isaiah 55:10–11

THE INSPIRATION OF SCRIPTURE

Behind the idea that the Bible is the "inspired Word of God" is a claim of authority. The Bible is authoritative or normative for Christians, but what that means has evolved over the centuries and varies in different Christian traditions. Both Galileo and Bellarmine assumed the authority of Scripture, a book that was closely associated with God's *very* word; after all, if God "said it," that is pretty authoritative! I want to briefly explore this topic because the term *inspiration* is so vague it allows conversations to take place between two Christians, who both believe in the Bible is "inspired," but who are nevertheless talking at complete cross purposes. Since this leads to misunderstandings between Christians over theological issues, it is only to be expected that such confusion spreads into discussions between theology and other disciplines, such as science.

The Old Testament prophets gave oracles that were deemed by the people to be inspired by God. Over time that inspiration became applied to the whole canon of Scripture (the Greek *canon* means "measuring stick," i.e., a reference standard). Instead of a human agent being inspired, the notion of divine inspiration was transferred to an entire *book*. Consequently, the biblical text was—and is—deemed authoritative and the source of God's revealed truth. However, the obvious danger with over-venerating the Bible as "God's Word" is the sin of idolatry. There is only one Word, the *logos*, and that is Jesus Christ—to whom the Scriptures bear witness (John 1:1, 14; Heb 1:1–2).

This is not the place to look at the history of how the Bible came to be put together to form the canon of Scripture.[1] But that was very much a complex human process, however much Christians may want to claim that it was also guided by the Holy Spirit. Those who formulated the canon were acknowledging that the books involved were *already* recognized as authoritative by the various Christian communities around the Mediterranean, from Alexandria to Rome. Nevertheless there were additional books that some communities took to be inspired that, for one reason or another, were not finally recognized as such by the church councils.[2] This illustrates that the notion of "inspiration" on its own was not a sufficient criterion for inclusion within the canon. The *content* of those books had also to be deemed coherent with the received teaching of the apostles, convey the true representation of God and of his saving acts, and provided trustworthy guidance for the needs of the community.[3] Consequently, and as Bird concludes: "Scripture is both a product of tradition and a part of the church's ongoing tradition, and it cannot be interpreted as a document of faith apart from that context of communal interpretation and use."[4]

As we saw in the 1616 decision, the implicit understanding of the inspiration of Scripture meant that the Holy Spirit, in effect, dictated the text to the individual authors. As such there is an unequivocal claim that God is the true author of Scripture. If this is the case—and if we also believe that God directed the process of forming the canon—then clearly there can

---

1. For further details see Harrington, "Introduction to the Canon," 7–21, Evans and Tov, *Exploring the Origins of the Bible*, and Bruce, *Canon of Scripture*.

2. For example: *The Shepherd of Hermas*, *The Gospel of Peter*, *The Apocalypse of Peter*, *Acts of Paul*, *Barnabas*, *The Gospel of Philip*, and several other works are not part of the New Testament.

3. Bird, "Authority of the Bible," 48–49.

4. Ibid., 63.

be no internal contradictions or inconsistencies between different parts of the texts, either within one book or the whole of Scripture. The advantage of this position is that it creates a degree of certitude when considering the Bible. If we find one part of a particular book to be somewhat obscure then we can legitimately look for clarification in another text (deemed to be on the same topic) since, ultimately, God is the author of everything. (Although this begs the question, if God is the ultimate author why is there some obscurity in the first place!) In this approach you consider the *whole* canon as integrated; interpreting the Old Testament in the light of the New, and interpreting a difficult passage with one that is deemed to be clear(er). The degree to which an individual or Christian tradition emphasizes *God's* authorship correlates with the elevation of the biblical text, and the certainty that is attributed to it.

But what was the role of the human author? Did God literally dictate the text to them in a manner that Muslims traditionally believe God did in communicating the Qur'an to Mohammed? If this is the case then the human authors, in essence, had their fallible humanity temporarily suspended while writing the text to create something that is—literally—error free. And the word "inerrant" is often connected with Scripture in some traditions. The effect of this is to elevate Scripture as being beyond reproach, above question, and simply to be accepted as it is. We saw this in Bellarmine's stance in the previous chapter. This approach seems to be untenable on many levels, for what happened to the free will that God graciously gave humankind at creation? Why is it *necessary* for God to suspend the author's humanity, so to speak, simply to communicate the messages that God gives them? Is that consistent with the character of God Christians claim is actually revealed in Scripture? And why, when you look at the Bible as a whole, does it not have the consistency you might expect from a single author?[5]

A more basic question is simply: "*Who* says that the Bible is the inspired word of God?" Some might want to answer that emphatically: "God does"! However that cannot be the case. While some of the Old Testament prophets use the proclamation "Thus sayeth the LORD" to announce an oracle (which always sounds more authoritative when using the traditional King James version of the Bible!), in general, no biblical author claims inspiration from God in writing their particular text. Perhaps the one that

---

5. Moreover, there are many genres to biblical literature including historical narrative, poetry, prophetic oracles, wisdom literature, gospel accounts, letters, and apocalyptic literature. The biblical text is clearly not homogeneous in its literary style and this differentiation in itself is very important in terms of biblical interpretation.

comes nearest is the often-quoted verse which says: "All Scripture is inspired by God" (2 Tim 3:16). But to *what* does that text refer? At that time it meant the *Jewish* Scriptures and recognizing them as also important and authoritative for the Christian church. In those days that meant the Greek Septuagint (or LXX) translation that was in wide circulation across the Hellenized world. Those Scriptures were divided into three sections: the Law of Moses, the Prophets, and the Writings. There was also a hierarchy within the canon, with the Pentateuch (the Law or Torah) having preeminence. The Sadducees and Pharisees had different canons, with the Sadducees only recognizing the Law as authoritative.[6] In some parts of the New Testament the Jewish Scriptures are referred to simply as "the Law and the Prophets" (e.g., Matt 5:17; 22:20; Luke 16:16; 24:27; and Acts 28:23). Only in Luke 24:44 is there reference to a three-part canon: "the law of Moses, the prophets, and the psalms."[7] The author of 2 Timothy is therefore endorsing the view that *all* sections of the Jewish Scriptures were deemed to be authoritative for Christians and not just one or two of the three elements. If we, as later interpreters, extend that to include the New Testament as well, then clearly that claim is of human origin, not divine.

The point I am stressing, one that is widely recognized but does not get stated explicitly often enough, is that it is *humankind* that claims that the Bible is the inspired word of God. *We* make that assertion; it is a *confessional* statement. That being the case, we can always ask the legitimate question: "On what grounds do we make that claim?"

A confessional statement is proclaimed by a *faith community*. People see a certain text as inspired because its contents resonate with their own experience of the divine in such profound ways that the community recognizes that the author's text goes beyond his immediate audience.[8] It is in

---

6. Ibid., 44.

7. Moreover, we can see the way Matthew, Paul, and other New Testament writers make use of the Old Testament Scriptures and *reinterpret* them for their context. For example, the way Matthew's uses Old Testament prophecies, which are arguably the most explicit communications of God, and adapts them in his birth narrative demonstrates that even prophetic oracles were not fossilized or seen as the unchangeable word of God. What Matthew, and the other gospel writers, did was to interpret the life, death, and resurrection of Jesus in the light of the Jewish Scriptures. See also Enns, *Inspiration and Incarnation*, 113–63.

8. When you go to an art gallery and you see a picture that somehow speaks to your heart you may claim that the artist was inspired simply because of the impact that painting has upon you. When a scientist comes up with an original idea that is game changing within his or her discipline, someone else may marvel and say that they were inspired.

this sense that a later community asserts that God must have inspired the original author because his message is still timely and insightful to another generation of believers. Consequently, the debate over which books should be in the canon is something that is decided by the believing community itself. There is an essential principle here: a community asserts the inspiration of the biblical text. And there is an important corollary: if you're not part of that community you will not find it inspired.

Another point to recognize is that the biblical authors were writing their text to specific situations. Or later redactors (editors) were combining texts and editing them in such a way as to address *their* audiences. This means that the author (and/or redactor) had an agenda or a purpose in writing (or editing and arranging) the text as they did. In doing so they were implicitly addressing the needs and the questions of their community at a specific point in time—for example, a community returning from exile and seeking to reestablish their identity. The questions and issues in their context are not the ones that we have today and therefore one must be cautious in simply transplanting those messages for our world and congregations. As we will see in the next section, we approach the task of interpretation with a sense of responsibility and humility, rather than with an arrogant spirit of certainty. And ultimately it is our faith communities that will decide whether through that process the Holy Spirit has spoken to them today. Indeed it is only *they* that can say, "That was an inspired sermon," not the preacher!

THE INTERPRETATION OF SCRIPTURE

I have already mentioned, albeit very briefly, the development of literal and nonliteral interpretations in the patristic period, and how that developed further in the Middle Ages. A new age of scholarship emerged in the Renaissance with its thirst for ancient sources in their original languages. One result was that Erasmus of Rotterdam published a new Greek New Testament in 1516 which challenged the longstanding Latin Vulgate. Around the same time Martin Luther maintained a strong emphasis on biblical authority, primarily because the gospels pointed to Jesus Christ and to God's redemptive activity in the world. Luther was also noted for advocating sacred Scripture as its own interpreter, and consequently each passage of Scripture should be interpreted in terms of the theology of the whole canon. Even inconsistencies in Scripture didn't trouble him because they did not touch the heart of the gospel.[9] Moreover, there was also a "fuller sense" to Scripture, meaning that contemporary interpretations and "applications" of Scripture were also intended *by God* even if they were not consciously intended by the original authors.[10] Calvin, who like Erasmus was influenced by Humanism, strived to understand the historical context of the original biblical texts before applying them to the issues of his day. And, as we have seen, Calvin emphasized the language of accommodation; in revelation, God condescends to our linguistic and cognitive level. Even so, both reformers rejected fanciful allegorical readings of the text and adopted a more straightforward literal-historical approach.

Regardless of differences in interpretation, the Bible throughout history up to that point was viewed, essentially, as a divine book. In other words, God as the ultimate author was never seriously questioned. This implies that any one verse of Scripture is as equally God-inspired as any other.[11] Galileo and Bellarmine would have agreed on this point and, on this matter, the early reformers agreed with the Roman Catholic Church. This framework, or paradigm, was significantly challenged during the Enlightenment.

With the Age of Reason came what is referred to as the historical-critical method, which is still very influential today. The Enlightenment rejected traditional sources of authority, such as Scripture, monarchy, and the church, and replaced them with the use of reason and the scientific method. Consequently historical-grammatical criticism treats the Bible like any other book of antiquity. This approach was—and is—enhanced by the study of linguistics, ancient languages and literature, together with history, archeology, and other related academic disciplines. Since this method has no theological commitment to orthodoxy, Scripture is not regarded as divinely authored. Indeed, such a claim is deemed an irrelevance in studying the text. This new paradigm was therefore seen as an unprejudiced, or

9. For example, in addition to chronological inconsistencies, the New Testament writers do not cite the Old Testament prophets verbatim. Calvin held the same view: see González, "Bible in Christian Tradition," 102.

10. Merold Westphal points out that "application," a concept often linked to preaching, is misleading if it implies "the transition from theory to practice. The movement is from then to now. Pauline epistles often have a first 'half' of heavy 'theology,' followed by a second 'half' of 'ethical' exhortations and guidance. The hermeneutical task in *both* cases is to hear what God is saying to us now, in different contexts, through what human authors said to their readership then." Westphal, "Philosophical/Theological View," 85, his emphasis.

11. This approach lends itself to "proof texting."

"scientific," approach to study historical works. Consequently, any differences between various ancient manuscripts became an object of fascination and debate. Once what was deemed to be the most reliable (composite) Greek New Testament (based on available ancient fragments and manuscripts) had been established by broad consensus, then further study of the text's *contents* could begin. Again any differences and similarities (say, between the writers of the Synoptic Gospels) became a topic of further scrutiny and debate as to their significance.[12] Moreover, this approach examined the historical, cultural, social, and political context of each biblical book. For example, since each gospel account was written to a different audience at diverse historical and geographical locations, what additional insights might that give the modern reader? The historical criticism approach, which from the outset paid little attention to theological interpretation, inevitably challenged the traditional, single (divine) author view of Scripture. The human authors were given greater freedom and responsibility in the way that they collected and presented their material. Consequently, the assumed uniformity of God's revelation throughout the whole of the Bible was also challenged.

Since historical criticism emerged alongside the development of modern science, it paralleled science's assumptions of its own methodology. In particular, in the absolute "objectivity" of the scholar along with a complete detachment from the object of study—the biblical text, in this case, rather than a natural phenomenon. Knowledge about both "objects" was greatly enhanced by such methodologies, as they are by modern scholarship today. The rapid progress of Western science, technology, economic wealth, and political power led to a general air of cultural optimism. Enlightened humankind, now free from the shackles of traditional authority, was deemed to be advancing forward toward a new, sophisticated, secular civilization. This fantasy was finally shattered by World War I, whose devastation was the antithesis of a progressive, enlightened society. This collapse of confidence fueled further introspection—the seeds of which were already evident to the discerning long before. I will briefly explore this "crisis" in the context of biblical interpretation as this watershed fueled new post-modern paradigm(s).

12. Matthew and Luke are commonly read as revised and expanded versions of the Gospel of Mark, which is widely deemed to have been written first. Look, for example, at the gospel writers' perspective of Jesus's teaching on divorce in all three Synoptic Gospels and compare the similarities and differences (Mark 10:11–12; Matt 5:31–32; 19:9; Luke 16:18).

Bradley McLean approaches hermeneutics from a philosophical perspective and begins by clarifying two ways in which a text can have meaning. The first is the *founding* sense-event which contains the triplet of (i) the original social and historical context, (ii) the author's intentions and belief system, and (iii) language itself. The second form of meaning is the *present* sense-event, which is the significance of the founding sense-event for us today. This illustrates that there can be two types of meaning for a biblical text, the original founding sense-event and another that reinterprets the significance of that event in every subsequent generation. What is problematic for hermeneutics is the self-limitation of the historical critical method, which always sees the observer as detached from the object. This means that its methodology actually *excludes* the very possibility of the present sense-event.[13]

Friedrich Schleiermacher (1768–1834), like the early Protestant reformers, maintained that the biblical texts had an original meaning, but one that had been lost over time. He was primarily interested in the *author's intent* and claimed this could be discovered by scrutiny of the original languages of Scripture and a rigorous linguistic analysis of the text. In contrast to the early reformers, he was not so interested in the theological unity of Scripture, but in the differences and the historical particularity of the texts.[14]

A problem arises, however: how can you understand an individual element of a text without knowing its *context* within the whole of the text? You have to read the complete text before you can understand or appreciate its constituent parts. The interpretive process is therefore circular and iterative. A general understanding of the whole text shapes how you interpret the individual parts within it, and vice versa. As Merold Westphal writes, this whole process is very complex:

> There are two major circles for Schleiermacher: one is grammatical-linguistic, the other is psychological. In the first case, the movement from part to ever-larger whole goes from sentence (1) to text (this periscope, this chapter, this book), (2) to genre (or textual tradition), (3) to the whole language shared by the author and original readers, (4) and finally to the history of human language. In the second case, one moves from this work of the author (1) to the author's entire body of writing, (2) to the author's whole life as

---

13. McLean, *Biblical Interpretation and Philosophical Hermeneutics*, 31.
14. Ibid., 37.

known to us from other sources, (3) and finally to what we know of the nation and era to which the author belongs. The first circle focusses on the text, the second on the author.[15]

This, then, is Schleiermacher's hermeneutical approach, which he considered to be more of an art than a science since it required skill, patience, good judgment, and even intuition, rather than a mere technique. It is nevertheless scholarly. The study of founding sense-events is therefore not merely a reading of the biblical text, but involves linguistics, literature, history, archeology, etc., as the investigator endeavors to immerse his/herself into biblical times and worlds, and into the minds of the authors.

Schleiermacher's emphasis on authorial intent has a legacy that lives on today. Knowing the author of a biblical book or letter is perceived, in some way, as authenticating the text and stabilizing its meaning. This is, in part, why the debate over Paul's authorship of the contested letters continues.[16] It's not about their place in the canon that is being questioned (if it is to some, then there is too strong a link between canonicity and authorship); rather it's about the overall shape of Paul's theology (with respect to the undisputed letters). This also builds upon the idea that the *authors* determine the textual meaning, and this continues to be the commonsense approach of interpreting the Bible. However there are many texts whose authors are unknown; one obvious example is the letter/sermon to the Hebrews. But even the four canonical gospels themselves are all anonymous and *attributed* to Matthew, Mark, Luke, and John, respectively. Anonymity creates a problem when strong emphasis is placed on the quest for the author's intent. This is relaxed if one allows for a "fuller sense" of Scripture or if one considers the continuing significance of the text in present sense-events.

As a related aside, consider the sayings of Jesus in the gospels. The physical presence of Jesus within the gospels is made more vivid by the spoken words that the gospel writers attribute to Jesus. The effect of such speech is to make Jesus the author of the gospel, rather than the gospel writers themselves. This is most evident when we say, unconsciously, "Jesus said . . . ," rather than "the gospel writer tells us that Jesus said. . . . " Notice also how even the use of "tells us" is speech language that makes the writer more present, whereas using the word "wrote" makes the absence of the author more apparent. This is not meant to be pedantic or cast doubt as

---

15. Westphal, *Whose Community? Which Interpretation?*, 28.

16. These are: Colossians, Ephesians, 2 Thessalonians, and the Pastoral Letters—1 & 2 Timothy and Titus.

to whether Jesus really said a particular statement; rather it demonstrates how the physical presence of the person, implied by speech, is deemed to be more significant than what was merely written. The gospel writers were, like Paul, proficient in rhetorical skills and wrote persuasively![17] Those, then, who today use a "red-letter Bible" to highlight "the very words of Jesus" have their biblical interpretation influenced by Romanticism.

Roland Barthes (1915–1980) famously proclaimed the "death of the author" by which he challenged the way some people (including Schleiermacher) appeal to the author—and hence authorial intent—as a means of controlling or limiting the meaning of texts. Instead, Barthes saw the text's sense—or significance—as more open-ended and not relying so much on the author but on the text's later interpreters (i.e., present sense-events). Consequently, this need not necessarily be a rejection of the "author," since it is good to be as informed as one can be about the text. However, the "Paul" who scholars ultimately create by their skillful research will necessarily always be a construct and not the "real" Paul.[18] There will always be an unknowability of the real author and so we should not place excessive weight on our constructs, insightful though they can be. Barthes was therefore critiquing Schleiermacher's confidence in the "knowability" of the author from studying the text and its context. McLean adds that "from a theological perspective, the doctrine of the inspiration of the whole of Scripture does not require that we resuscitate historical authors to guarantee the truth of its message."[19]

Another famous "death" is Friedrich Nietzsche's "death of God" and, moreover, that *we* have killed him! In the context of the Enlightenment, knowledge was seen as a commodity determined by "objective" human reason. If "truth" depends upon the historical context of the culture that purports to determine the truth, then this "truth" is continually being revised. This means there is no privileged position or perspective and, consequently, *any* truth can be proclaimed as *the* Truth. Not only is this the foundation of relativism, but since human rationality now determines that truth, God's role as the ultimate foundation of truth has been usurped. God is therefore "dead." Nietzsche's conclusion was that the overall effect of the

---

17. See Witherington, *New Testament Rhetoric*.

18. The same is true of the quest for the "historical" Jesus. These constructs tend to say more about the different presuppositions of the scholars involved, which is why there can be such controversy over "Paul," "Jesus," etc.

19. McLean, *Biblical Interpretation and Philosophical Hermeneutics*, 54.

Enlightenment program has been to empty human existence of any essential meaning, purpose, truth, or value—leading to nihilism.[20]

Nietzsche's *nihil* ("emptiness" or "nothing") reflects a crisis of significance and shakes the foundations of historicism, which naively saw the historian as an unbiased, detached "subject" and history as an "object" *out there* to be discovered and reconstructed from objective "facts" from "sources." This crisis of significance affects all related disciplines—including biblical studies—with the corresponding loss of conviction in relating the conclusions from the historical-critical method to contemporary issues, like ethics.[21] Nihilism challenges the inherent confidence of modernism and opens up the way for postmodern perspectives. McLean concludes: "The notion of objective biblical interpretation is a myth of the Enlightenment. It is not a theological principle, or a principle of the Reformation, or even a biblical principle. It is simply a myth."[22]

The crisis of significance also led to a crisis in the assumed way of knowing—in the very rationality of the scholar. Whatever a person knows about history, the Bible, indeed *anything*, depends on the perspective of the investigator and his/her acts of interpretation. Since every interpreter is located in a social and historical context, the interpretation must therefore be limited by the worldview of the interpreter. Michel Foucault expressed the view that the assumed autonomy of the subject, as being outside and above history, is itself a historical construct/concept that is rooted in the Enlightenment.[23] The impact of this was a further unraveling of confidence, such as in the assumption of intellectual *progress*, the underlying belief that there was a single, linear story-line—or "metanarrative"—to history, and this led to a general disenchantment. This is what lies behind the emergence of postmodernism.

Historicism's philosophical foundations over the nature of the "subject" (historian, biblical scholar, etc.) and the "object" (history, biblical text, etc.) have been undermined, leaving a monument that is in ruins. Nevertheless,

---

20. Ibid., 65. See also Plantinga et al., *Introduction to Christian Theology*, 95.

21. This "crisis of significance" can also be argued as an inevitable consequence historicism's methodology which rules out present-sense events and only considers founding-sense events. If a methodology begins with "the only questions worth asking are these kinds of questions" then we can't complain when we desire the answers (or responses) to questions that are outside that initial remit. One does not ask the scientist of the significance of a kiss with the expectation of a meaningful response!

22. McLean, *Biblical Interpretation and Philosophical Hermeneutics*, 73.

23. Ibid., 81.

McLean rightly argues that historicism is a "splendid ruin." For this reason historicism continues to have an honored place within a post-critical hermeneutics paradigm.[24] Since the notions of subjecthood and reason are being questioned it is perhaps no surprise that a postmodern response is explored within the context of existentialism, which explicitly brings the interpreter into the way of knowing.

The story of historicism's "ruin" parallels a similar subject-object dualism which also suffered a fatal blow. This arose in the philosophy of science with the birth of quantum mechanics, which seriously undermined the assumed independent status of the "observer" of classical physics. Interestingly, much of the physical sciences continue as before, oblivious to this crisis of knowing that birthed the new paradigm of modern physics. It is not that classical physics is "wrong" *per se*, indeed much of the success of science and engineering is founded on classical physics. Rather, the absolute confidence that its methodology presupposed has been undermined by the nature of the quantum world and the necessary involvement of the hitherto detached "observer." The implications of the new paradigm were uncomfortable to the founders of modern physics, like Einstein and Schrodinger, and are still being wrestled with today.[25] Both classical physics and historical criticism still have an important role to play, but the total confidence in the rule of reason that founded these paradigms has gone.

Rudolph Bultmann (1884–1976)—and Karl Barth (1886–1968)—wrestled with the crisis of historicism. Bultmann, influenced by the existentialism of Martin Heidegger (1889–1976), distanced himself from the constructed "historical Jesus." He argued: if our faith is closely bound to such a rational reconstruction, then it is essentially rooted in unbelief. For Bultmann, ultimately, the "Jesus of history" is irrelevant to the "Christ of faith."[26] He also posed an interesting question: "Is biblical exegesis without presuppositions possible?"[27] The answer was—and is: no! All scholars approach a text with presuppositions, or prior assumptions about how that text should be read. No scholar, however self-aware, could be objective and neutral; we all read Scripture through a "lens." Those spectacles not only

24. Ibid., 93–94.

25. This will be explored further in chapter 5.

26. Ibid., 148. There are parallels with Tertullian's "What has Athens got to do with Jerusalem?"

27. For further discussion, see Silva, "Contemporary Theories of Biblical Interpretation," 109–11, and McLean, *Biblical Interpretation and Philosophical Hermeneutics*, 143–56.

include our theology, but also our culture, social standing, educational heritage, gender, race, politics, etc. All these factors define our "location" and this means that we cannot approach the text without a bias. There is *always* an element of subjectivity. This is not a "bad" thing to overcome, rather something of which we should be acutely self-aware. What is therefore necessary is that the interpreter raises his/her presuppositions to the *conscious* level and allows them to be tested, so putting them at risk. Authentic engagement with the text occurs when you allow the text to challenge and change your starting assumptions.

If we step back and reflect on our experience of knowing, we can recognize this important—and often overlooked—principle: faith *precedes* knowledge. We need faith in our presuppositions, or our knowledge foundations, before (and after) we build upon them—whether this is in the context of historicism or science.[28] This is reminiscent of St. Anselm's famous dictum: "I do not seek to understand in order that I may believe, but I believe in order to understand." St. Augustine had a similar saying: "Unless you believe you will not understand." Having faith requires a *commitment prior* to the outcome of the engagement with the biblical text (or the scientific experiment). This act is the sign that the interpreter is not detached but intimately involved in the interpretation process. It is curious that this notion from a patristic father and a medieval scholar, who clearly were not influenced by the Enlightenment, still seems relevant for today in light of the intellectual "crises" discussed.[29] Perhaps they were inspired!

In light of the crisis of historicism, how does one now approach biblical interpretation? Schleiermacher's hermeneutical circle was all about *knowledge* at both the micro and macro levels. However, it did not take into account the inevitable presuppositions of the interpreter, shaped by his or her contingent social and historical location. As mentioned earlier, a philosophical hermeneutic that incorporates the interpreter inevitably entails an existential—or experiential—perspective. It's not about developing a new technique of knowing, one that strives to overcome our bias—since pure "objectivity" is impossible. Rather, we allow the text to make us consciously aware of—and challenge—our own presuppositions and tradition, and to affect change. In this new hermeneutical circle, we read the text

---

28. Newbigin writes: "Both faith and doubt have their proper roles in the whole enterprise of knowing, but faith is primary and doubt is secondary because rational doubt depends upon beliefs that sustain our doubt." Newbigin, *Proper Confidence*, 105.

29. This is also an example of a present sense-event, reinterpreting a founding sense-event for today.

(and its context) and the text reads us (and our context); it is an open-ended "dialogue." For example, how do we react when we read the gospel accounts of Jesus's encounters with the woman caught in adultery (John 8:1–11), the Canaanite woman (Matt 15:21–28), or the rich young ruler (Mark 10:17–25)? These texts will evoke different responses depending on our gender, ethnicity, social and economic status, and religious tradition. Consequently, this emphasizes interpretation as a present-sense event, not the founding-sense event, as the process probes the significance of the text on *ourselves* as the interpreters. This inevitably means there are *multiple* present-sense events, not just one "interpretation" to which we must all assent. This, therefore, embraces the "subjective" element of interpretation, which is now "relative" rather than "universal." The degree to which we are self-aware and allow the text to "read" us, so causing us to be transformed in the process, demonstrates our authenticity in the activity of interpretation.

I have mentioned that our tradition, as well as ourselves, are open to being challenged and changed in this revised hermeneutical circle. "Tradition" in this context means the particular religious community (e.g., Anglican, Roman Catholic, Pentecostal, etc.) that shapes our worldview, and this framework is itself embedded in political, social, and cultural structures.[30] This tradition functions as a "lens" through which we view Scripture. Our tradition may also be associated with more modern perspectives to which we self-identify, like feminism, liberation theology, ethnicity, sexual orientation, etc., as well as our political allegiances. The quest for authenticity requires us to be open to the presuppositions of our tradition *also* being critiqued, as they have covert or overt "power" over the texts and how we presently read them. We should have the courage to question: "Is this what the text is saying to me, or is that what my tradition says the text is saying?" This requires us to both examine our tradition and be thoroughly informed of other traditions and their viewpoints. This critique is not to focus on suspicion, or the fear of being deceived by our own tradition; rather it is a positive means for hope by which we encounter a new reign of God's Truth—which itself is a work of the Spirit.

---

30. For those in the West, this includes: capitalism, democracy, individualism, freedoms and rights of citizenship, secularism, materialism, and an Enlightenment-shaped educational system.

PHILOSOPHY AND THEOLOGY

So much for postmodern *philosophical* hermeneutics, you may exclaim, but what about *theology* and Christian devotion? It is true that the historical critical method was—and is—a means to study the biblical text without recourse to ecclesiastical interference, but that does not necessarily imply an antagonism toward theology or the church. On the other hand, it is quite understandable that Christians study the Bible with different motives and expectations. This can be as a means or expression of personal piety—for reflection, encouragement, and guidance. Alternatively, the Bible can be read by a pastor with the view to give a sermon, or by a professor who is teaching biblical studies at a seminary. Regardless of the motivation, the Bible is inevitably being "interpreted" and we now recognize that there is much more to interpretation than one might first think.

As we have seen, philosophical hermeneutics is general and applies in the engagement of any text, not just the Bible. We can, as part of a tradition, also incorporate a *theological* lens of our own construction.[31] What is required, however, is an awareness of that lens's existence, its assumptions, and the way it might restrict or control our interpretation of the text. Moreover, while that theological lens may be a defining feature of our tradition (i.e., our faith community), it is in itself *not* divine. We assent to that lens, consciously or unconsciously.

Consider that theological lens for a moment as we return to the notion of the Bible as an inspired text. Those who claim a "top-down" authority to Scripture often, as a consequence, view the Bible as propositional truth (i.e., conveying broad, factual information).[32] This can be challenging when various parts of Scripture are in contradiction, or—to use a more neutral term—exhibit "diversity."[33] Rather than questioning the "Bible as

---

31. See, e.g., Migliore, *Faith Seeking Understanding*, 50–63.

32. By "top-down" authority, I mean that the biblical canon is self-authenticating and/or is divinely authored; either way the Bible is deemed "infallible" or "inerrant."

33. One, perhaps esoteric but not insignificant, example relates to Holy Week. The Gospel of John's chronology differs from the Synoptic Gospels in that—for John—Passover was on Saturday, rather than (Good) Friday. The symbolic parallels with Passover are then different. In the Synoptic Gospels the connection is with the Passover meal and the institution of Holy Communion. In John the symbolism links the death of Jesus with the slaughter of the Passover lambs. See O'Day, "John," 704–5, 719, 814. Other more significant differences or diversity are present, too. For example, the histories related in Samuel and Kings are different from those in Chronicles. From the "top-down," single-author perspective how is this to be understood? See Enns, *Inspiration and Incarnation*,

propositional truth" presupposition, all efforts are focused on resolving the apparent conflict. Moreover, the assumed understanding of what divine authorship entails means that the conflict should, in principle, be resolvable. If it cannot be resolved, this might lead to a crisis of faith when, in fact, it should be regarded as a crisis of presuppositions. It is not the Christian faith *per se* that is in crisis, but our faith in our dearly held assumptions as to the grounds of that faith. On the other hand, those who take a "bottom-up" approach to biblical inspiration (i.e., it is the believing faith community that authenticates Scripture) do not have the same problem. The role of the human authors and their different social and cultural locations, and their diverse audiences, provides legitimate space to circumvent these difficulties.[34] The key point, then, is to do with the *character* of Scripture. While many want to claim Bible as having, or containing, a propositional nature; others recognize its personal character and that Scripture therefore emphasizes relationships and witness. The latter has a subjective or experiential element to it and so can be readily reconciled with a postmodern framework; the former strives for objectivity and hence is associated with the paradigm of modernity. Yet both camps can claim the Bible as an "inspired" text.

Following McLean, I have stated that there can be two different aspects, or emphases, in biblical interpretation: founding sense-events and present sense-events. To reiterate, a founding sense-event is traditionally studied using historical-grammatical *exegesis*, where you focus on a detailed and critical analysis of the text, the original language, and the original historical situation. Consequently the goal was to bring *out* what the original author meant and his context, i.e., *exegesis*. If this is all a minister does in his or her sermons, one can rightly ask the question of relevance: "What does this mean—or what is God's message—for *today*, in our context?" Alternatively, if a preacher focuses on present sense-events, the danger is an overemphasis on the contemporary interpreter and our context. This can lead to *eisegesis* whereby the interpreter reads *into* the text his or her own bias and presuppositions. A sermon may have lots of contemporary relevance, but have little connection to the biblical text and *its* context. In such cases the biblical text is merely a pretext for the preacher to say what they wish.

This brief overview of philosophical hermeneutics has shown that historicism's noble exegetical goal is, ultimately, unattainable. We can never

---

71–111.

34. See the discussion in Holladay, "Reading the Bible," 130–31. This human factor can be augmented by the theological perspective of progressive revelation.

truly understand the original author's intent, or fully grasp the audience's understanding of the text. But this does not mean that authorial meaning can be simply ignored and so "anything goes" in our interpretation. Nevertheless, being open to the present sense-events can empower interpretations that others would rather not sanction. This has occurred in the past, whereby Scripture has been used to justify racism and slavery, or to promote a "prosperity gospel." There is, therefore, an *ethical* dimension to the use and abuse of Scripture of which we must be aware. An exegetical study of the founding sense event, as best as we are able, at least acts as a kind of "guardrail" to present sense-events. Wise and sensitive preachers feel the weight of this responsibility as they wrestle with interpreting the text for today.

Returning briefly to the role of theology in hermeneutics, in addition to having an explicitly Christian lens as part of our tradition, some Christians see value in *canonical criticism* as a counter-balance to, yet embedded within, the paradigm of historical criticism. Canonical criticism is explicitly theological, yet it takes seriously the insights gained by historical criticism and other critical methodologies. Nevertheless, it emphasizes the text's final form and the shape of the whole canon. For example, a question that canonical criticism considers is: "What is the theological significance in the way the New Testament books are arranged—beginning with the Gospels, rather than, say, in the chronological order of when the texts were written?" In so doing, canonical criticism seeks to see coherence in God's revelation within the texts as a whole.[35] Furthermore, canonical criticism also sees the location of that revelation as being in the believing communities that shaped the canon and who understood the canon as normative for faith and practice.[36] Canonical criticism, then, enables overarching biblical themes from creation to the eschaton to be explored, and hence a linear, metanarrative to be affirmed—from a *theological* perspective.

Another way of thinking about this matter is to reconsider the "top-down" and "bottom-up" descriptors of the inspiration of Scripture. An exclusive "top-down" approach leads to divine dictation and making an idol of Scripture. An exclusive "bottom-up" approach, as taken in the historical critical method, makes the biblical text an aspect of anthropology. Only a

---

35. This should not be misunderstood as simply putting the clock back five hundred years, thereby ignoring historical criticism and reasserting the Bible as "God's word" by ecclesiastical dogma.

36. Ibid., 134–35. See also Wall, "Canonical View," 111–30, and Childs, *Biblical Theology of the Old and New Testaments*.

believing faith community, by means of a confessional statement, can claim aspects of *both*. Our recognition of the Bible's resonance with the human condition throughout history and across many cultures leads us to assert the authors were divinely inspired. Yet our theological understanding of God, derived from the Scriptures—and iterating through the hermeneutical circle—informs us that the process of divine inspiration does not override the free will of the authors graciously given by God in creation.

THE PURPOSE OF SCRIPTURE

The reason why this excursion into the inspiration and interpretation of Scripture was necessary is simple: *it is very significant in the context of science and faith*. Those who see the two in conflict often view the character, inspiration, and interpretation of Scripture very differently from those who don't. As we saw in the previous chapter, there are different views on the *intention* of Scripture. From within the framework of canonical criticism it is quite legitimate to endeavor to respond to the question: "What is the purpose of Scripture?" I take the view that Bible's primary end is to point to *God's salvific acts in history*.[37] God's actions reveal Godself and the Divine's purposes, a process that has evolved throughout Scripture. For example, we read of God's providence for Joseph in Egypt, for Moses and the exodus, and in enabling the Israelites's return from the Babylonian exile. God's initiation of saving acts is culminated in the life, death, and resurrection of Jesus—which extends to the *whole* world—and in the sending of the Holy Spirit. This is not just all the *peoples* of the world, fulfilling the initial Abrahamic blessing, but for the whole created order. I could add that one secondary objective is to reveal right ways for living as the people of God. We see this from the giving of the Ten Commandments to Moses, followed by the persistent call of the prophets' for repentance, to the Law's radical reinterpretation by Jesus in the Sermon on the Mount. Loosely speaking, this is consistent with "faith and morals" from the previous chapter. However,

---

37. This is consistent with the traditional confessions, such as the Belgic Confession (1561, 1619), The 39 Articles of Religion (1571), and Westminster Confession (1646–47), which all emphasize that the Bible contains all things needed *for salvation*. (see Schaff, *Creeds of Christendom*.) I appreciate that these early Protestant documents were more concerned with emphasizing the contents of Holy Scripture as *sufficient* for salvation, so distancing themselves from the authority of popes, church tradition, and the abusive practices of indulgences. Nevertheless, although there is a direct link between Scripture *and* salvation, there no mention of natural philosophy in these documents.

based on the principle of accommodation, I do not think that the Holy Spirit had any intention to address what we today call "scientific questions" when inspiring the biblical writers. Rather, and returning to 2 Timothy 3:14–17, the *purpose of Scripture is very clear*. James Dunn is emphatic:

> The text is clear: The sacredness of the writings is directed to the end of "making wise for salvation"; the point of Scripture's inspiration was that the Scriptures should be beneficial for teaching and equipping the student believer for effective living as a Christian. Since this text is the most explicit biblical statement of what Scripture is *for*, the fact that it targets the purpose of Scripture so explicitly, and with a clearly delimited scope, should be given more weight, both in the doctrine and the use of Scripture. Too much time is misspent asking of Scripture what it was not designed to answer. Better that Scripture itself should instruct us as to what its purpose is.[38]

As we will see in the chapter 4, this viewpoint is foundational in the context of dialogue between science and Christian faith. Having laid the biblical foundations for Christianity, it is time to scrutinize the basis of science and its methodology.

---

38. Dunn, "2 Timothy," 853, his emphasis.

Chapter 3

On the Nature of Science

A Philosopher is a person who knows less and less about more and more, until he knows nothing about everything. A Scientist is a person who knows more and more about less and less, until he knows everything about nothing. —John M. Ziman[1]

Philosophy begins in wonder. And, at the end, when philosophic thought has done its best, the wonder remains. —Alfred North Whitehead[2]

INTRODUCTION

At present there is a significant discrepancy between the views in some sectors of the general public concerning the reality of human involvement in climate change and the safety of genetically modified foods, and the views of scientists researching in those areas. In contrast, people are generally enthralled and inspired by whatever astronomers, cosmologists, and particle physicists have to say! Society's attitude toward the authority of science is variable and changing in our increasingly postmodern world. Nevertheless, we begin this section with the more traditional view of unquestioned confidence in scientific knowledge, perhaps to the point of mystically

---

1. Ziman, *Force of Knowledge*, 119.
2. Whitehead, *Modes of Thought*, 168.

accepting the validity of the findings of science, whose inner workings are now deemed too technical for the nonexpert to grasp.

The common view is that science leads to proven, universal knowledge, derived in a rigorous way from the facts of experience acquired by impartial observation and repeatable experiments. Only science can produce unprejudiced, reliable knowledge. And this is far superior to any personal opinions—especially any conclusions derived with reference to an unseen God. While this portrayal may be overexaggerated, I think the description, for the most part, still resonates with popular perception. This view is accepted on two grounds. First, knowledge of nature, via science, has enabled humankind to manipulate and control nature in unprecedented ways resulting in ongoing industrial, technological, and medical revolutions. Science is stupendously successful! And, second, because the classic "scientific method" is naively perceived to be true. However, "the end justifies the means" is philosophically unsatisfying and should always be suspect. We need to be self-critical of "truth" claims—lest we create a new form of dogma, or, as Nietzsche put it, a "will to power."

Science's supremacy as *the* way of knowing has, however, not always been the case, as Stanesby points out:

> In the Middle Ages theology was described as the "Queen of the Sciences," that is, the highest and most authoritative form of knowledge. All rational enquiry had to conform to the canons of theological thought.... The religious view of the world dominated all thinking, and whenever there were clashes the religious view won the day.... Today natural science rules as queen over all and is commonly accepted as the supreme source of all knowledge.... The tables have been turned. *Contemporary religious thinkers now tend to take the authority of science for granted and try to match their theology to the prevailing Western scientific tradition.*[3]

Today, the bottom line is simply that when faith and science appear to conflict or confront each other, faith must give way to science. To suggest the reverse produces uncomfortable ridicule and is perceived as holding firm to groundless dogma. It even throws suspicion on one's intellectual credibility. Different ways of relating science to Christian faith will be explored in the next chapter; suffice to say here that conflict can only arise if both are claiming to answer the *same* questions. As a complement to the last chapter,

---

3. Stanesby, *Science, Reason and Religion*, 1–2, emphasis mine.

this chapter presents a very brief historical outline on the nature of science and to explore its "authority."

WHAT IS SCIENCE?

What elements are needed for a scientific investigation?[4] First, we need a healthy respect for the material world—for all that is in the universe—as a *worthy object of study*. This may be obvious for most in the Western world, but historically it is important because Platonic thought considered direct experimentation on the natural world to be of little value, preferring instead to describe it in terms of nonphysical "ideals." One such preconceived ideal was that, as the (planetary) gods were sublime, the motion of heavenly bodies *necessarily* had to follow the perfect form, namely, circles.[5] In addition to the different views of the Greek philosophers, the other prevailing worldviews of the past were generally with respect to polytheistic gods (e.g., Egyptian, Babylonian, Greek, and Roman) or various forms of monotheism (Jewish, Islamic, and Christian). Both religious *and* philosophical biases can inhibit experimental investigation or try to constrain a view of the universe due to prior assumptions or beliefs. Religious influence, however, is not always detrimental—as the history of science demonstrates.[6] Eventually people began to recognize that if God thought matter was worth creating, then it might be beneficial to actually study what was made. This was—and is—endorsed in the apparent reliability, harmony, and order of nature. The observed repeatable patterns within nature led to mathematical abstraction and the search for basic mechanisms which provided a "cause and effect" connection between its components. In brief, the scientist strives, ideally, to observe the world *as it is* and not to be prejudiced by some external authority or self-imposed notions. On one hand this is liberating, but on the other, nature is what it is, and we cannot force it to fit into our mold.

4. The name "science" comes from the Latin word *scientia* which means "knowledge."

5. When inconsistencies were observed, little circles were added to model mathematically the planet's motion so creating the appearance of "wheels-within-wheels." This picture was perpetuated until Kepler showed that planetary motion was "better" described by an ellipse. Ironically, Kepler (who was aware of the Apollonius's work on conic sections) hesitated to adopt elliptical orbits thinking: if it were as simple as that then the problem would have already been solved by Archimedes and Apollonius. ["Conic sections," well-known to Greek and Roman mathematicians, are all part of the same geometrical family, consisting of: circles, ellipses, parabolas and hyperbolas.]

6. See, e.g., Hooykaas, *Religion and Modern Science*.

There should be openness to fresh evidence but in conjunction with the test of experience.

Second, scientists also generally assume the universe to have *objective reality*, be *intelligible*, and be *uniform*. All three presuppositions are essential, although we often take them for granted and fail to appreciate their importance. We will not enter the world of *The Matrix*, or the dream worlds of *Inception*! Rather we will assert that the universe really exists for all of us and is not a product of our brain's imagination; in that sense it has an objective reality. We cannot strictly prove this, although we all generally live our lives on the basis that it is true! Second, the intelligibility of the universe is, I think, surprisingly profound; Einstein commented: "The most incomprehensible thing about the universe is that it is comprehensible." Finally, uniformity simply means the regularity and processes identified on a limited scale are assumed to be valid throughout the universe.

Presuppositions, by their very nature *cannot* be proven; they are taken as a "given" or "on trust." Building on the above three foundations can therefore be said to be an exercise of "faith." This is true for all scientists, regardless of our personal religious or philosophical viewpoint. The Renaissance scientists all had a Christian perspective, which provided them with confidence in the objective reality of the universe and its ultimate intelligibility. They perceived themselves as "thinking God's thoughts after him"; the existence of the Creator God was their guarantor of rationality. Although many scientists today would reject the religious views of the Renaissance scientists, we still rely on the above three fundamental assumptions to practice our profession. As highlighted by René Descartes, the issue of trusting presuppositions is a fundamental problem that will not go away. It needs to be addressed seriously in whatever philosophical (or religious) position the scientist chooses to adopt. It is also worth emphasizing that science itself is not the property, or under the jurisdiction, of a specific worldview; not atheism, not Christianity, not Islam, and not that of ancient Greece. It is simply a way of studying nature. The *methodology* of science, however, strives for unified self-consistent mechanisms within a closed universe, and can be regarded as rationalistic in character.[7] But to be a scientist still requires *faith* in the validity of science's presuppositions.

---

7. A "closed universe," in this context, is one where recourse to God as an explanation or cause is excluded from a *scientific* explanation. Invoking God in this way is unproductive for science and, perhaps to the surprise of some, counterproductive for Christianity. This is referred to as the "God-of-the-gaps," i.e., whenever we don't understand something in nature we simply claim "God did it"! The problem with this stance is

The Scientific Method

How is scientific knowledge acquired? How do you define what is science, as opposed to pseudo-science? Often people speak of a "scientific method" as a *prescription* for obtaining new knowledge. This was first clearly stated by Francis Bacon (1561–1626) and, in modern terms, the scientific method consists of the following procedures:

1. Observation and classification of relevant "facts."

2. Generalization by means of inductive reasoning.

3. Construction of a theoretical framework that allows one to make deductions.

4. Verification of the predictions by experiment.

This traditional scientific method sounds familiar, logical, and plausible. However, it can be criticized in two distinct ways.[8] The first is more philosophical: will this process, in fact, lead to reliable, universal knowledge? The second is historical: with four hundred years' worth of hindsight, has science actually followed this procedure and has knowledge of nature developed in this way? We will consider both perspectives critically in this chapter.

First consider the nature of "observation," which is ultimately a sense experience by a conscious being.[9] When two or more observers view an object to what extent can we say that they see the same thing? In addition to the existence of color blindness, we are aware of what we call optical illusions. In the latter, the same physical "information" has reached the retina but it is interpreted in different ways by different people. (Can you "see" the

---

that it leads to a diminishing role for God as science progresses and provides a reasonable explanation for that previously mysterious "gap." Instead, theoretical chemist Charles Coulson exhorts: "When we come to the scientifically unknown, our correct policy is not to rejoice because we have found God; it is to become better scientists" (cited in Polkinghorne, *One World*, 60).

8. I acknowledge Alan Chalmers for the following critique: see Chalmers, *What Is This Thing Called Science?* Similar critiques can be found in most introductory texts to the philosophy of science. Another short review can be found in Polkinghorne, *One World*, 6–25.

9. "Observation" was initially visual but now includes *indirect* methods, like the use of the telescope or microscope, and other more modern devices (oscilloscopes, transducers, computers, etc.). In such cases, these indirect "observations" are eventually converted into visual representations (or some other sense perception). The word "observation" is retained for simplicity.

two different images in each of the illustrations shown in fig. 1?) Perception, therefore, is not *uniquely* determined by the image on the retina, but it also depends on past experience and culture, as well as our personal knowledge, training, and expectations. In addition, we are aware of the famous "moon illusion" where a full moon appears significantly larger near the horizon than it does higher up in the sky. These examples serve to demonstrate that the real nature of observation is more subjective than the scientific method presupposes. The "facts" of our sense experience are not always to be trusted.

Figure 1a

Figure 1b.

Second, "inductive reasoning" is a logical process by which a *general* statement is made on the basis of a finite number of observations. A simple example to illustrate this process begins with the observation that on a certain day and time the planet Mars appeared at a specific position in the night sky. The truth of this *particular* statement can be established by any number of careful observers, using their senses.[10] Kepler performed a systematic study of the red planet's motion and concluded that its path around the sun had an elliptical shape. By studying the other planets, Kepler generalized his finding and stated that all planets will move in ellipses around their sun. This general assertion is a *universal* statement, but it is based on past experience and it is only applicable to similar systems.[11] The process of deriving a universal generalization from particular statements of our own experience is called the principle of *induction*. It is also clear that to make a universal statement with any degree of confidence requires the number of observations to be large, repeated under a variety of conditions, and no reliable observation should be in conflict with the derived generalization.

Is, however, the principle of induction reliable? Does it lead to conclusive knowledge? This question has been well studied and the principle of induction has been shown to be flawed. The problem is that from a set of true observations (particular statements) you cannot prove logically that the general statement will *necessarily* be true. The premises may well be true but you can be led to a false conclusion. Just because all the swans that you have observed are white does not logically prove that all swans are, in fact, white. Moreover, even if the principle of induction has worked well on a previous occasion, we cannot infer that induction *always* works. That is unacceptable because we are endeavoring to use the principle of induction to prove the principle of induction! This seriously undermines confidence in the traditional scientific method; any claims that arise from induction are only *probably* true but no absolute claim to "proof" is valid.

Furthermore, practical difficulties arise too: How many observations are enough to justify confident induction? What is meant by observations under a "wide variety" of circumstances? These problems arise from the vagueness in the description of the scientific method, which appeals to "reasonableness." Yet we can be deceived by reasonable claims that we consider

---

10. This is precisely what the Jesuit astronomers did in "confirming" Galileo's observations.

11. This assumes each planet is *only* interacting with the sun and not some additional, nearby massive planet—like Jupiter.

to be obvious or self-evident. At one time it was obvious and reasonable to consider the earth to be stationary and at the center of the universe—that was, and still is, after all, our sensory experience!

Chalmers points out another practical issue that, to me as an experimental physicist, is highly pertinent, namely: general statements are often in the form of an *exact* mathematical equation. However, the particular statements from observation that provide the evidence for that generalization will inevitably have a degree of inexactness about them. The data points will always have some statistical scatter about the form of the generalized, exact mathematical function. Much more could be said here, but Chalmers's essential point is simple: "It is difficult to see how *exact* laws can ever be justified on the basis of *inexact* evidence."[12]

Third, the scientific method assumes the existence of "facts" that are "out there" and pregnant with meaning. Indeed, the word "data" is loaded with implications of inherent significance—unlike the word "blip." Such facts are meant to be prior to and independent from any conceptual framework or theory. But how can a scientist recognize from the outset that what is reaching us *is* intrinsically meaningful, that data *is* data? How can we sort out from the myriad of external sense stimuli that *these* particular ones are the "signal" and the rest is just "noise"? Obviously, we have to know what to look for and where to look, that this "fact" is significant and the rest can be ignored. That filtering process requires *training* of some kind so that you have a conceptual framework to classify *relevant* facts. How else can a medical specialist recognize the presence of a disease from an X-ray or an MRI image? Our minds are not, as Locke insisted, a blank tablet on which the senses write. Observation itself is not without reference to anything else, i.e., a self-evident fact from which I can proceed to generalize free from any preconceived ideas.[13] Rather facts are always interpreted facts; observation is built upon some sort of preexisting theory. Stephen Toulmin cautions:

> The structure of a scientific theory may be built up entirely from the bricks of observation, but the exact position the bricks occupy

---

12. Chalmers, *What Is This Thing Called Science?*, 50, emphasis mine.

13. It is our preconceived ideas that allow us to recognize that the famous constructions of the graphic artist M. C. Escher (1898–1972) are "impossible." In addition to the tentativeness of our initial conceptual framework, which may itself be mistaken (e.g., Aristotelian physics of motion), the particular statements themselves (i.e., our recognition of the "facts" by observation) are not infallible simply because we are finite human beings. Scientists make honest mistakes.

depends on the layout of the scientist's conceptual scaffolding; and this element of scaffolding, which the scientist introduces himself, is always open to misinterpretation.[14]

We can endeavor to qualify the scientific method to address these criticisms. But try as we may, we cannot separate observation from its theoretical framework. In the end, we cannot throw out the idealized concept of observation entirely, but we must acknowledge that observations are built on presuppositions which, in turn, have limitations that need to be recognized.

Perhaps surprisingly, all this sounds remarkably similar to Schleiermacher's hermeneutical circle from the previous chapter. There, you will recall, was a basic problem of interpretation: how can a scholar understand an individual element of a text without knowing its context within the whole of the text? Schleiermacher's hermeneutical circle was a circular and iterative process, where the study of the complete text was required before you could understand or appreciate its constituent parts, and vice versa. There is always more to discover, whether that be for the historian, biblical scholar, or theologian; so it is for the scientist. Scientists are still compiling our book of nature, and the more we know of this beautiful complex universe, we are humbled—or should be—by how little we do know. As with historical criticism, the induction principle is valuable to science and its theories can cautiously develop using it. But the idea that rigorous, provable scientific knowledge is provided by this route has to be abandoned. It is an ideal that cannot be substantiated in principle and therefore, if applied, it must be interpreted with strict caution. This conclusion seems unsatisfactory in a discipline where human reason and knowledge are paramount. Is there a better approach?

Falsificationism

Sometimes when you are wrestling with an intractable problem, it may be that you are asking the wrong question. Instead the issue needs to be reframed or seen from a radically different perspective. One such example is Karl Popper's alternative approach to the inductive method, based on "falsification."[15] This method acknowledges that observation is guided by

14. Toulmin, *Return to Cosmology*, 26.

15. Falsification is discussed at length in Chalmers, *What Is This Thing Called Science?*, 59–103.

and presupposes theory. A scientific investigation starts with "problems"—and a problem is only recognized as such in the light of some preexisting framework. Moreover, Popper was not concerned with "proof" or trying to establish a particular theory as "true"—notions we often traditionally associate with science.[16] Instead, a theory was simply to be viewed as a tentative formulation created to give an adequate account of some phenomenon, which has needed to be devised to overcome problems encountered in previous theories. Once proposed, the new theory should be subjected to a rigorous range of experimental tests. If the theory fails to stand up to the tests, then it must be eliminated and an alternate replacement be devised. In this description, science progresses by trial and error, by conjectures and refutations: only the fittest theories will survive. Consequently, a theory cannot be proven "true"; instead it can only be said to be "confirmed" as the best available to date.

But not any old theory will be appropriate for this approach; the major condition is that the proposed hypothesis be actually *falsifiable*. This means that there must be some possible observation or experiment that could in principle refute the theory. The assertion "It always rains on Thanksgiving weekends" can be tested and so the statement is falsifiable. But the statement "All bachelors are unmarried" is necessarily true by definition of the word bachelor, and so it is inherently not falsifiable. These kinds of unfalsifiable statements should have no place in founding scientific theories. This may seem pedantic, but Popper wanted to reject some areas of what we currently call science on these grounds. Certain aspects of social science and psychology can be criticized as containing claims that cannot in principle be falsified.

One feature of "falsificationism" is a sense of scientific knowledge *progressing* as promising theories flourish while inferior ones are weeded out. A promising theory has special merit if it (a) has a wide range of applicability, (b) is full of specific detail, and (c) is bold in its novel predictions. Clearly, (a) and (b) create more opportunities of showing it to be in error. A theory that excels in both these aspects and yet survives all the stringent tests is a theory of significance. But an additional desirable quality is that of *boldness*. Bold or courageous theories clash with the currently accepted scientific thinking and this deserves some merit. Without this quality, science

---

16. The historical context was that of Logical Positivism, whose influence still pervades society even though it has been discredited. See also Godfrey-Smith, *Theory and Reality*, 19–37.

would cease to progress into new fields. Furthermore, the "confirmation" of a bold theory is particularly important as it inevitably must falsify some part of the accepted background knowledge with respect to which the conjecture is termed "bold." If the refuted knowledge was/is central to the overall foundations of science *at the time*, the repercussions can be very serious indeed. Newton's law of gravitation (1686), Maxwell's description of the electromagnetic field (1864), Einstein's general relativity (1915) are just three examples of bold theories which eventually led to novel predictions which progressed science in significant ways. Chalmers's summarizes by saying that, in contrast to inductivism, falsification makes no claims

> to the effect that the survival of tests shows a theory to be true or probably true. At best, the results of such tests show a theory to be an improvement on its predecessor. *The falsificationist settles for progress rather than truth.*[17]

Karl Popper echoes the same sentiment:

> The empirical basis of objective science has thus nothing "absolute" about it. Science does not rest on solid bedrock. The bold structure of its theories rises, as it were, above the swamp. It is like a building erected on piles. The piles are driven down from above into the swamp, but not down into any natural or "given" base; and if we stop driving the piles deeper, it is not because we have reached firm ground. We simply stop when we are satisfied that the piles are firm enough to carry the structure, at least for the time being.[18]

Knowledge obtained by science, according to falsificationism, is therefore always tentative.

The story does not end there, however, as there is a serious *logical* critique of falsificationism based on the fallibility of observations. All observations are fallible (as they are for inductivism), yet crucially for falsificationism, a theory must be rejected when it fails a test when, possibly, it is the test's actual observation that is in error. Within this methodology, there is no obvious mechanism for clearly distinguishing between a wrong (but genuine) observation and a wrong theory. Any proposed mechanism would inevitably have a degree of subjectivity associated with it: you cannot establish which of the two is wrong without comparison to a third reference.

---

17. Chalmers, *What Is This Thing Called Science?*, 86, emphasis mine.
18. Popper, *Logic of Scientific Discovery*, 93–94.

And how can you objectively validate the third reference? Moreover, if the theory is not confirmed by the result of the test, the theory cannot be falsified conclusively because you cannot rule out the possibility that some part of the complex test situation, other than the theory under test, is responsible for the outcome.[19] The historical example of Tycho Brahe's claim to have refuted the Copernican theory illustrates this point.

The Danish astronomer correctly argued that *if* the earth orbited the sun, then the direction in which a nearby star is observed from the earth should vary during the course of the year, with respect to distant stars, as the earth moved from one side of the sun to the other. But when Brahe tried to detect this predicted stellar parallax with his instruments, which were the most accurate and sensitive ones in existence at the time, he failed. This led Brahe to conclude that the Copernican theory was false. With hindsight, it can be appreciated that it was not the Copernican theory that was responsible for the faulty prediction, but one of Brahe auxiliary assumptions.[20] Brahe's estimate of the distance to the fixed stars was many times too small. When his estimate is replaced by a more realistic one, the predicted parallax turns out to be too small to be detected by Brahe's instruments.

Had falsification's own critical criteria been adhered to, the Copernican theory—and many of the best examples of scientific theories—would have been rejected outright in their infancy. But they were not. In the end, Popper himself admits that it is often necessary to retain theories despite apparent falsifications.[21] In the light of the inherent and persistent difficulties in falsificationism and inductivism perhaps we should take a more serious look at *history* in order to see if the actual development of science can enlighten the situation.

---

19. This is the Duhem-Quine Thesis; for further discussion in the context of monotheism, see McGrath, *Science and Religion*, 67–71.

20. This example also demonstrates that to make a specific "test" usually involves accepting further numerous assumptions, including those theories relating to the physical instruments that we are using. These auxiliary assumptions require additional justification to provide some degree of confidence that the result of the principle test was not just an experimental artifact or some manifestation of another physical phenomenon.

21. Chalmers, *What Is This Thing Called Science?*, 103.

The Revolutionary and Evolutionary Development of Scientific Knowledge

The history of the development of science has been analyzed by many, but a landmark review was undertaken by Kuhn in the 1960s, which will be outlined briefly as it reveals important new insights. His analysis shows that science hasn't always grown continuously in a gradual, evolutionary manner; rather, steady progress is interspersed with times of dramatic change and rapid development. He summarized the progress of science in the following repetitive, open-ended scheme:

Pre-science—Normal Science—Crisis/Revolution—New Normal Science—New Crisis . . .

"Pre-science" is a diverse collection of disjointed, embryonic ideas that eventually become structured and coherent. The resulting framework is adopted by the scientific community and provides a "stable" environment in which a scientific discipline can undergo its routine problem-solving activities. This general scientific worldview, or *paradigm* as Kuhn called it, is a network of presuppositions, theories and techniques that is required to established Kuhn's phase of "normal science." An example is the group of theories (e.g., Newton's mechanics, wave optics, and Maxwell's electromagnetism) that constitute "classical" physics. The success of classical mechanics led to the idea of a clockwork universe. Determinism was then absorbed into the presuppositions of the paradigm, so much so that it was unthinkable that nature could be anything other than a clockwork system of cause and effect.[22]

In this "normal science" regime, scientists assume that the paradigm provides all the means for solving any problem that it faces. A failure to solve a problem is seen as a failure of the scientist rather than the paradigm. Stubborn difficulties are seen as serious *anomalies*, rather than the falsification of the underlying worldview. Consequently, strict falsification is rejected, as there will always be some phenomena that cannot be understood within the current paradigm. An example of an anomaly would be Brahe's discovery that the widely observed comet of 1577 traveled through the supposed crystal spheres separating the planets. Although this challenged the Aristotelian perspective of immutability in the superlunary region, it was, by itself, not enough to overturn that longstanding paradigm. Neither was the fact that planetary motion was observed not to be perfectly circular;

---

22. Quantum mechanics was to challenge that, as we will discuss in chapter 5.

Ptolemy and Copernicus used additional circles (epicycles) to overcome that anomaly. Further stunning evidence came with the observation of the 1604 supernova—a very rare phenomenon. The astronomers of the day lived (uneasily) with the nova anomaly but, still, the whole of Aristotelian physics was not disbanded simply because of this one falsification. However with all these anomalies taken altogether, combined with Kepler's ellipses and Galileo's observations, a gradual acceptance of the heliocentric perspective—or paradigm—began. This new paradigm profoundly changed the way we view the cosmos. This irreversible discontinuity with the precious worldview is a feature of a "paradigm shift."

A "crisis" arises, as we have just seen, if enough anomalies are encountered that scientists themselves become uneasy and start questioning the fundamentals. A sign of serious conflict with the reigning paradigm is when scientists propose more radical, even bizarre, theoretical solutions in an attempt to the resolve the conflict. Philosophical disputes can arise as scientists reexamine the presuppositions of the paradigm. Some of these assumptions may have received little scrutiny before because they are so absorbed into the worldview that they seem reasonable to all. In addition to the presuppositions of Aristotelian physics, a more modern example was the assumed eternal, uniform nature of time as a stage on which the play of the universe is enacted—a presupposition that Einstein's theory of relativity challenged.

Even when something serious is considered widely to be wrong with the prevailing paradigm it is still extremely difficult to break out and view the problem from a totally new perspective. This is simply because we have unconsciously absorbed so much of the current worldview. Our own fallacies are often the hardest to spot, just because they are our own. Scientific genii are born when they make great strides in resolving, in a coherent manner, those troublesome anomalies and introduce new frameworks that, in time, become paradigms (e.g., Copernicus's heliocentricity, classical physics, and quantum mechanics) which profoundly change the way we view the world. But the genii behind scientific revolutions also struggled with the implications of the novel theories they devised, despite their mathematical success. The relevant scientific community's "conversion" to the new paradigm is also a contentious and stressful process for all its practitioners. However, once a new paradigm is widely perceived to be established, then normal science can again continue, albeit in a new light—and until the next crisis develops.

Kuhn's depiction of scientific development is not without criticism. Is this merely *descriptive* or is it meant to be more? Falsification inherently had science as *progressing* within its methodology; in what sense is science progressing for Kuhn?[23] Is Kuhn saying that scientific knowledge is therefore *relative*—dependent on the values of the community who validate their paradigm? Is there really a clean discontinuity with the previous paradigm, or can you live in two overlapping worldviews?

In considering that last question for a moment; it could be said that the introduction of relativity and quantum mechanics just showed a regime in which the mathematics of classical physics was not valid. It is as if the scientific worldview was like a map. It was not that the classical map was wrong everywhere, rather new explorations at the (then) edges showed it to be inadequate—those "edges" being at speeds approaching that of light and the miniature world of atoms and their constituents. In this sense the old paradigm is not entirely abandoned once a new one is deemed necessary. But this qualification to Kuhn's paradigm is not entirely fair. It may be true in a pragmatic sense, i.e., for all practical, everyday purposes, but both relativity and quantum mechanics challenged irreversibly the philosophical basis for the classical worldview. These philosophical aspects (e.g., Determinism) are rejected; the purely clockwork universe is dead. But it dies slowly, particularly as quantum mechanics is so counterintuitive; it is difficult to grasp its concepts let alone its implications for philosophy.

The history of the development of science has been—and still is—a controversial subject. It is often a matter of opinion as to the significance of certain contextual features (like politics, economics, society, ideology, and religion) that shaped the scientists who actually had the original ideas, and their communities. Feyerabend proposed, controversially, that the history of science demonstrates there is no ahistorical, universal scientific way to knowing, and suggested—in effect—"anything goes." Perhaps there is a false binary that is being juxtaposed: either there is a method (in which case, what precisely is it?) or there is not (so embrace relativism or anarchy). Is there a middle way—one that allows science's methodology to be more open-ended, one that progresses along with scientific knowledge? Or is this just idealistic self-denial? Not surprisingly, there is lack of universal agreement on this matter!

---

23. That emphasis on *advancement* can be connected with Enlightenment's own worldview.

Despite all this scrutiny and skepticism, what we do know from experience is that science "works." Yet, however much we want to know *why* the scientific enterprise is so effective, we have not been able to establish rigorously a rational basis for our confidence in science. Many are uncomfortable to be reduced to appealing to the ill-defined notion of "common sense"; that seems like admitting intellectual defeat. In the end though, it seems that perhaps the straightforward conclusion is to recognize that science is simply a *human* enterprise. An apparently successful enterprise—one undertaken by a community—but a human endeavor nonetheless. The fantastic achievements of science encourage and empower humankind to regard ourselves as autonomous or *above* nature. A closer look at humankind's negative achievements, enabled by science, suggests—at the very least—we should be more humble. It is too easy to focus on the benefits to society that science has enabled and overlook growing ethical problems of pollution, wealth disparity, and the killing capability of weapons—to name but a few—all of which have been fed by the sciences' mastery over nature. The scientific endeavor contains the brilliance of the human intellect, along with the inherent fallibility of being human. That weakness also includes recognizing the difference between knowledge and wisdom. The trouble with the success of classical physics, and of the determinism and reductionism within its worldview, is that we have forgotten what it means to be human. Science had absolute faith in the objectivity of the observer and the power of reason. We forget that science not only contains logical deductions, but interpretation, inspiration, intuition, and skill—all human qualities—all easily appreciated in an artistic endeavor.

PERSONAL KNOWLEDGE AND COMMUNAL AUTHORITY

Michael Polanyi made an important contribution to the philosophy of science, one that seems to be often overlooked—yet one that made a significant impact on theologians, such as Lesslie Newbigin. One can see why from McGrath:

> Polanyi's fundamental assertion is that all knowledge—whether it relates to the natural sciences, religion or philosophy—is *personal* in nature. Polanyi's post-critical approach to the nature of knowledge argues that knowledge must involve personal commitment. Although knowledge involves concepts or ideas, it also involves

something more profound—a personal involvement with that which is known.[24]

Some of Polanyi's ideas have already been mentioned implicitly above in critiquing the scientific method, especially in the nature of the observer. Rather than detached objectivity there is instead personal commitment, something that rings true for me and accurately describes the scientist's passionate involvement in both knowing and the known.

Scientists not only use their senses but also use tools that are purpose-built for the study at hand. This is analogous to the way that surgeons use their instruments, or a carpenter uses a hammer, or a person who is blind uses a white stick, or someone reading uses spectacles. They are not focusing their attention on the tool but on the object that the tool is manipulating or sensing. If you focus on holding the hammer you are likely to miss the nail that you are trying to hit! Those tools, therefore, become an extension of their bodies such that the person *indwells* the instrument. For scientists, that indwelling involves implicit trust in our instruments and our perceptions in the study of nature. While we use our instruments, we do so *a-critically*; we cannot at the same time rely on it *and* doubt it.[25] There is, therefore, an existential element to knowing. It is not just our minds that are involved; our senses, augmented by the instruments we indwell, are also intimately and actively a part of knowing.

Recollect from the previous chapter the existential element to post-critical hermeneutics. Polanyi's ideas resonate strongly with our experience of interpretation; that is why science and faith are much closer than people sometimes think. Both involve acts of interpretation, in both cases knowledge is personal and requires a commitment to the means of knowing and the known.

As mentioned earlier in the context of Kuhn's paradigms, science functions within a community and that tradition authorizes science itself. The practice of science involves sponsorship and publication of results, both of which require the approval of one's peers. Consider the process of publishing a scientific paper: the personal knowledge of the scientist now

---

24. McGrath, *Science and Religion*, 84–85.

25. Newbigin, *Gospel in a Pluralist Society*, 34. This does not mean that scientists use their instruments naively. Intense scrutiny over the inexactness that the instrument itself introduces into the observation, along with a thorough understanding of all the auxiliary assumptions and principles of the instrument and its use, is undertaken by every skillful experimentalist. Even with this rigor, honest mistakes can still occur and that in itself is a learning process that progresses scientific knowledge.

becomes shared public knowledge, a process not without risk since papers can be rejected. The scientist's personal judgment is being scrutinized by others with appropriate training and experience. Without the discoverer's findings being authenticated by his or her peer community there is a sense that scientist's findings remain "private truth," to borrow Newbigin's phrase. However, the scientist's personal knowledge is not merely subjective but has universal intent. The scientific community's endorsement is the means by which private truth becomes "public truth." This peer-review process does not guarantee the veracity of the public truth; there is always a sense in which knowledge is tentative and open to future revision. We have all heard in the news of scientific stories of dramatic discoveries which were later retracted because no one else could repeat the observations. In addition, in what is now recognized with hindsight as the beginning of one of Kuhn's crisis phases, the private truths of Copernicus and Galileo were actually correct, even if the tradition of their day refused to accept their findings as public truth. Although these two illustrative examples are embarrassing or awkward for the cause of science, it is the *community* that, in time, provides the self-correction—which again emphasizes the role of the scientific tradition itself in authenticating science.

The parallels for the Christian—of both being a part of a larger community (and its important role in stabilizing interpretation) and of passionate personal knowledge and commitment—are evident. It is for this reason that this highly selective—and inevitably personal—historical review of the philosophy of science is presented here. Having surveyed the topics of the nature of biblical interpretation and the nature of science we are now in a better position to examine ways in which science and faith can relate to each other. Even if you disagree with me, by being explicit on these matters, we have a basis on which we can communicate. From these two foundations, either—or both—of which are often overlooked, or based on misleading perceptions, we can begin to have a meaningful conversation.

Chapter 4

On Ways of Relating Science and Christianity

> Let no man out of a weak conceit of sobriety, or an ill-applied moderation, think or maintain, that a man can search too far or be too well studied in the book of God's word, or in the book of God's works; divinity or [natural] philosophy; but rather let men endeavor an endless progress or proficience [sic] in both.
> —Francis Bacon, *Advancement of Learning* (1605)[1]

INTRODUCTION

What is the relationship between science and Christianity? Some regard the two as enemies, poles apart, and a battleground for the mind and soul. In contrast, others see a more symbiotic relation between the two; that science and faith can interact to their mutual advantage. Still others see them as completely separate—one in the realm of faith and the other of reason. Finally, in contrast to isolation, some regard "all truth as God's truth" and consequently see science and faith as, ultimately, united.

---

1. This quote from Francis Bacon's *Advancement of Learning* is cited by Charles Darwin, immediately after his title page, in his *Origin of the Species*. I acknowledge David Wilkinson for bringing this to my attention; see Wilkinson, "Reading Genesis," 142.

On Ways of Relating Science and Christianity

Ian Barbour (1928–2013) suggested four different ways or models to characterize the relationship between science and Christian faith.[2] The categories are: Conflict, Independence, Dialogue and Integration. It may be that this scheme is too simplistic and insufficiently nuanced; that the relationship between science and religion is too complex, too dependent upon historical and cultural contexts to characterize their relationship in that simple way. Nevertheless, I suggest this scheme provides a valuable framework and starting point. Therefore, Barbour's four classifications will be outlined briefly, with comments on their strengths and weaknesses from a Christian perspective.

CONFLICT

If you adhere to a strictly literal interpretation of the Bible, then it is inevitable that you will not believe in theory of evolution (in any form: micro or macro), and consequently you will see a direct conflict between those who promote the gradual development of life from simple to complex organisms and the created order given by God's spoken command in Genesis 1. Unfortunately, to complicate matters, because science's methodology is from the outset explicitly without reference to God, some Christians attack science as if it were the same thing as attacking atheism. It is, however, important to remember that not all atheists are antagonistic toward those with religious views; indeed many are respectful.

One form of atheism is scientific materialism, which assumes that (a) a scientific approach is the only reliable route to knowledge, and (b) the universe can be explained purely in terms of physical material, like atoms and molecules, and their interactions via the forces of nature. The first assumption concerns the way of knowing, which has been addressed in the previous chapter, and the second defines what is real. Materialism, as the name implies, isn't so much interested in nonphysical notions of, say, beauty and love, but only in the physical world; only matter has fundamental reality. The conclusion is simple: only science can tell us the true nature of reality. And the implication is obvious; the scientific explanation is the only one of real value. This viewpoint is fading these days; we are in an age that is more skeptical—or, perhaps, cynical—and less enamored with

---

2. Barbour, *Religion and Science*, 77–105. This classic book is an excellent resource for a more detailed discussion on this topic. Its main ideas have been distilled into a more accessible form in Barbour, *When Science Meets Religion*.

science's traditional authority within modernism. Still, it is worth discussing since this view is one that continues to linger in people's consciousness, not least because it is sometimes implicit in science and nature programs on television.

One line of scientific investigation, one that is a feature of scientific materialism, takes a reductionist approach where an explanation of a system is solely derived from the properties of its constituents. Reductionism implies that biological systems can be explained in terms of genetics, genes in terms of complex chemistry, and those chemical reactions as redistributions of atoms and molecules, which in turn are explained as a change in the mutual configurations of ions and electrons, and so on. Complexity, then, is to be explained by the constituent parts and their mutual interactions, ultimately by the interplay of fundamental particles. The danger with reductionism is that it takes its conclusions too far. Instead of simply explaining a phenomenon, it is considered it to be explained *away*. To conclude that *this* explanation is the *only* one that matters is an unacceptable philosophy. An investigation along reductionist lines is a useful approach to a scientific enquiry, but it is not the only kind of study. There are processes that *only* arise in complex systems, which therefore cannot be analyzed in isolation. Irreducible complexity is common in the life sciences, such as consciousness, brain function, and self-organizing systems, as well as in the physical sciences, in weather systems and other examples of chaos. Some caution is needed here, though, as what seems *irreducibly* complex today may well be revised tomorrow.

Scientific materialism is misconceived and is becoming more widely recognized as such; but the idea that science *explains away* still pervades. From a Christian perspective, and perhaps especially for a biblical literalist, some of the most contentious of scientific explanations concern the nature of humankind.[3] For example consider the statement: humankind is part of the animal kingdom; this is true. But to claim that we are *nothing but* an animal is an overstatement. A similar argument can be made from the statement: a human being is a machine. We can examine the different biological functions of all the parts of the human body and admire its interwoven complexity; but it is a gross exaggeration to then conclude we are *nothing but* a machine. Taken to the extreme, a human being is a collection of salts

---

3. It is worth recalling from the previous chapter that a scientific "explanation" simply elucidates something which was previously unknown, or a mystery, in terms of more familiar concepts within the presently accepted paradigm.

and water; but we are more than our basic chemical constituents. If we are only what we are made from then, as matter and energy are equivalent (from Einstein's $E = mc^2$), you and I could be quite reasonably expressed as a number of kilowatt-hours! No one lives their lives on the basis that there is no inherent value in the ordering of matter. If they did, then personality, being, thought, social values, even life itself, are simply meaningless! What is the origin and merit of ideas, truth, bravery, or morals? If you think about it for a moment, the very idea annihilates itself, because from that presupposition how can its own viewpoint even exist or have meaning? Polkinghorne expresses it admirably:

> The reductionist programme in the end subverts itself. Ultimately it is suicidal . . . it destroys rationality. Thought is replaced by electro-chemical events. Two such events cannot confront each other in rational discourse. They are neither right nor wrong. They simply happen. If our mental life is nothing but the humming activity of an immensely complexly-connected computer-like brain, who is to say whether the programme running on the intricate machine is correct or not? . . . If we are caught in the reductionist trap we have no means of judging intellectual truth. The very assertions of the reductionist himself are nothing but blips in the neural network of his brain. . . . Quite frankly, that cannot be right and none of us believes it to be so.[4]

We must be careful not to be deceived by claims made with the apparent authority of science but go beyond science's remit. The findings of science are only part of the picture. As Appleyard points out:

> Science begins by saying that it can only answer *this* kind of question and ends by claiming that *these* are the only questions that can be asked. Once the implications and shallowness of this trick are realized, fully realized, science will be humbled and we shall be free to celebrate our selves again.[5]

Returning to the theme of conflict; what is true, however, is that both a scientific materialist and a biblical literalist will agree that you cannot believe in both God *and* evolution. I will return to the topic of evolution later, but for now it understandable that those who are passionately committed

---

4. Polkinghorne, *One World*, 92–93. He also cites E. L. Mascall's dry humor: "However sure the scientist might be that other people were only elaborate machines, his protocol contained an escape clause for himself." Ibid., 92.

5. Appleyard, *Understanding the Present*, 249.

to those two positions will agree that science and Christianity have clashing truth claims. They are in conflict because both sides are making rival statements about the *same* domain, namely, the history of nature, and we are told that we must choose between them. Polarization, where there is thought to be only two possible positions, makes for easy news stories. The more extreme those views, the less chance that there will be of any reconciliation or some common central ground. It is sad that this warfare stance is still perpetuated today within the secular media, some church traditions, and by the so-called New Atheists. This confrontational approach is well over a century old and is oversimplified, often based on misunderstandings on the inspiration and interpretation of Scripture and/or the nature of scientific knowledge. In addition, conflict can be exacerbated by a jaundiced misreading or rewriting of history, as in the case of the trial of Galileo. Scientific materialism has promoted a particular philosophical commitment as a scientific conclusion, and the biblical literalists have promoted a prescientific cosmology as if it was an essential part of the Christian faith.

In conclusion, within some Christian traditions there seems to be an impenetrable impasse between science and faith. Is this important? It is certainly not a central issue to Christianity, despite the heated rhetoric to the contrary that you hear occasionally. However, what saddens me greatly is that this issue can become an unnecessary stumbling block to someone on their faith journey toward the Giver of Life. It is unnecessary because it is about faith in our presuppositions, not about faith in Jesus Christ and the transforming power of the gospel. Because of this reason I, frankly, find the stalemate tiresome and outdated. Let's explore other more positive, God-honoring alternatives![6]

INDEPENDENCE

An alternative view is that science and religion exist in two totally separate compartments. According to this perspective, there can be no conflict because science and religion refer to different domains or aspects of reality. Scientific and religious assertions are expressed in very dissimilar types of language and therefore they do not compete, as they serve completely different functions in human life. For example, it has often been said that

---

6. For a further assessment of the damage to Evangelicalism from Christian Fundamentalism's view of science, see Noll, *Scandal of the Evangelical Mind*, 177–208, 229–33, and Noll, *Jesus Christ and the Life of the Mind*, 99–124.

science asks "how?" and deals with the so-called objective facts of nature. Religion, on the other hand, asks "why?" and so deals with values and ultimate meaning or purpose.[7] Conflicts, therefore, only arise when we ignore these distinctions. That is, when religious people make claims concerning the natural order, and when scientists go beyond the area of their expertise to promote philosophical pronouncements.

Many, if not most, Christians have the highest regard for Scripture without insisting on six-day creation as literal truth. Some see the worlds of faith and reason as being independent realms, and are therefore not threatened by the findings of science. Karl Barth, the great neo-orthodox theologian of the last century, also adhered to this view. Since God is transcendent, God can only be known through God's self-disclosure, via accommodation, and—ultimately—in Christ. Consequently, natural theology, which uses arguments from nature (e.g., apparent design) to provide evidence for and to infer information about God, was deemed suspicious as it is based on human reason.[8] For Barth, Christianity depends entirely upon the divine initiative, not on our discovery. The primary sphere of God's action was in *history*, not in *nature*.[9]

There is tremendous depth to this approach. It acknowledges the mystery of our faith that we celebrate in the Eucharist, or Holy Communion. It correlates well with Kierkegaard's existential "leap of faith," emphasizing the risky commitment *to* faith and the corresponding limits of reason. Blaise Pascal's profound reflection carries the same idea: "The heart has its reasons which reason knows not of; we know this in innumerable ways."[10] This emphasis on the otherness of religious *experience* is consistent with having two separate languages for reality, one for faith and the other for reason.

This being the case, scientists are therefore free to carry out their work without interference from theology, and vice versa. Science makes quantitative predictions that can be tested experimentally, whereas the Christian faith is inevitably based in the language of symbols and analogy because of God's utter transcendence. A further consequent feature of the independence model is that faith will never be contradicted by the findings of science or affected by future changes in scientific knowledge. On a

---

7. Barbour, *Religion and Science*, 86.
8. McGrath, *Science and Religion*, 40.
9. Barbour, *Religion and Science*, 85.
10. Pascal, *Mind on Fire*, 230.

historical note, if Christianity had not absorbed the Aristotelian worldview, the Galileo crisis would never have happened. Within independence there is no need to align a Christian worldview with that of science, or—for that matter—with that of today's prevailing culture. More positively, such compartmentalism allows the Christian to accept at face value all the tentative conclusions of science—including evolutionary biology—since they have no relevance to the life of faith. Evolution is, after all, about natural processes and does not relate to the ultimate "why" questions of purpose or origins. This is a perfectly reasonable and respectable view, and Barbour concludes that it is a good "first approximation."[11] Nevertheless, independence can rightly be critiqued. While confrontation between science and faith is necessarily avoided, so is the possibility of any constructive interaction. From a theological perspective, the emphasis of God's transcendence is at the cost of God's immanence *in* creation.[12] Christians assert that God is *always* involved in creation, not that at the beginning God made the cosmos and then did nothing else within creation until the incarnation. This view overlooks God's active involvement in *continuing* creation.

In addition, this compartmentalization helps create and nurture the dichotomy between what Newbigin calls "public truth" and "private truth."[13] The privatization of religious faith, or the relegation of spirituality to the individual whim, means that there can be no consensus or public truth concerning values, meaning, morals, or purpose. While recognizing the sensibilities of living in today's pluralistic society, especially the need for mutual tolerance and grace, this total separation of scientific (or secular) and religious worldviews creates a serious problem for Christians with a mind for mission. I suggest this aspect is an underappreciated drawback for Christian traditions that advocate independence. Ironically when it comes to intractable social problems, bishops and senior clergy want to speak out in the political arenas, and politicians want to speak about morals and values! While the influence and tone of such rhetoric varies in different countries, no one underestimates the power of the religious vote in the United States.

Furthermore, I suggest that Christians should also be uncomfortable with strict independence because it could imply that what really matters is the saving of (nonphysical) souls and not a restoration of God's good

---

11. Barbour, *Religion and Science*, 88.
12. We will explore this theme further in chapters 6 and 7.
13. This is a common theme in Newbigin's writings; see, e.g., Newbigin, *Truth to Tell*.

creation. This attitude, exemplified in "this world is not my home; I am just a passin' through," undervalues nature with its escapist eschatology. Surprising as it may seem, the concept of body-soul dualism is not something that is endorsed in the Bible itself. Jewish thought, which was of course absorbed by the early church, has a holistic view of personhood. In the Hebrew Scriptures the self is a unified activity of thinking, feeling, willing, and acting.[14] Soul, from the Greek word *psyche*, is a principle of life. It is not something we possess but we *are* a soul because we are a living being made for relationships.[15] Body-soul dualism has its origin in Greek thought and was, in various forms, a cultural feature of the Hellenized world into which the gospel came. For Plato, the immortal soul enters the human body and survives after death, and one can recognize that this idea has been absorbed into much of Christian thinking. It is but a small step to then claim that matter is evil and that death is the liberation of the entrapped soul from the body. While that view is part of Gnosticism, which was officially rejected by the church, vestiges of it continued within the Christian life as evidenced in certain forms of asceticism and in the sense that the soul is superior over—or more important than—the body. The separation of body and soul continued well into the Middle Ages, and the presence of an immortal soul served as the demarcation between humankind and other creatures. Similar dualisms persist today; for example, the separation of the mind from the body can be traced back to René Descartes. The mere existence of psychosomatic illnesses, however, implies that there is a much closer connection between the mind and the body than this dualism allows. Modern science rejects such dualisms, and I argue they are *not* an inherent feature of biblical Christianity.

It is easy to see how the body-soul dualism, whose language still pervades the thinking of many church traditions today, correlates well with independence. For example, it is not uncommon to distinguish between the work of Jesus as addressing people's physical and spiritual needs, as if they were completely separate. If we read the gospel accounts afresh through a Hebraic lens, it opens our eyes to Christ's activities and miracles as simply examples of Jesus making people *whole*, in every sense of the word.

To conclude, in contrast to the conflict perspective, it is easy to be sympathetic with the independence view. Yet its rigid compartmentalization is, I think, too restrictive a description of the human condition and

---

14. Barbour, *When Science Meets Religion*, 129.
15. Green, "Soul," 358–59.

our ways of knowing and being. Moreover, there is a danger of projecting such isolationalism onto God, with too much emphasis on transcendence and with a corresponding downplaying of divine immanence and God's providential care.

DIALOGUE

Rather than either conflict or independence, a more constructive relationship between science and religion is that of dialogue. Dialogue asserts that science and Christianity offer *complementary* perspectives on the same reality, with the two worlds *not* being mutually exclusive. Dialogue emphasizes the similarities in the presuppositions, methods and concepts, whereas independence emphasizes the differences. In order to undertake genuine dialogue both scientists and theologians must have *critical* reflection on their own expertise, while respecting the *integrity* of the other. The purpose of dialogue is to foster mutual understanding, not for one to persuade the other of the rightness of one's own position. Dialogue deliberately falls short of the goal of integration, discussed in the next section, which aims to merge elements of the content of science with that of theology. In this sense dialogue looks for commonality in the general principles of each discipline, whereas integration looks further into the specifics with the goal of harmonization.

One aspect of dialogue is a comparison of the presuppositions and methodologies of the two fields and an acknowledgment of the presence of similarities in their ways of knowing. Elements of this comparison have already been addressed in the previous two chapters, including Polanyi's personal commitment to knowing and the role of the community. I have suggested that, from a postmodern perspective, the ways of knowing in scientific communities and Christian traditions are not as far apart as many imagine. Some will find that statement contentious. I suggest, however, that the traditional quest for knowledge in both areas says something more about our desire for certainty and our dislike of living with the twins of faith and doubt. I also mentioned that three of science's presuppositions are nature's objective reality, intelligibility, and uniformity. It is not at the outset obvious why the universe must *necessarily* possess intelligible order and structure, rather than randomness and chaos. Moreover, why this apparent regularity and rationality is comprehensible to a carbon-based life-form on an obscure planet near the edge of the Milky Way galaxy. Again, I stress that science requires faith in human reason. Furthermore, although physicists marvel at the elegance, beauty, and even the simplicity of the mathematical formulas that describe the basic forces of nature, nevertheless, having a set of equations does not automatically result in a universe, or explain it away. Hawking mused:

> What is it that breathes fire into the equations and makes a universe for them to describe? The usual approach of science of constructing a mathematical model cannot answer the questions of why there should be a universe for the model to describe. Why does the universe go to all the bother of existing?[16]

Not only then can a universe that is intelligible not be taken for granted, but neither can its very existence. The question "Why is there something, rather than nothing?" is a quandary that is beyond science's horizon of knowing. There is a contingency to our existence and to the order we recognize in nature; things don't have to be this way, it could have been otherwise.[17] Indeed, when you look at the some of the specifics of our universe (which we will in the next section) we are faced with the mysterious issue of *contingency*, which some take as evidence for a Creator who, if nothing else, chose the initial conditions for our universe.

There are many other topics that can be explored in dialogue, such as (a) a reverence for nature, (b) complex systems and self-organization, (c) information content, (d) consciousness and brain function, and (e) causation and indeterminacy; but this is not the place for such a review.[18] The essential point is that dialogue recognizes complementary perspectives on the same reality. Einstein famously said: "Religion without science is blind, science without religion is lame." And Pope John Paul II said:

> Science can purify religion from error and superstition; religion can purify science from idolatry and false absolutes. Each can draw the other into a wider world, a world in which both can flourish."[19]

---

16. Hawking, *Brief History of Time*, 174.

17. This is the basis of the traditional Cosmological Argument, which is outlined in appendix 1.

18. For those who are interested, a good starting place is Barbour, *When Science Meets Religion*.

19. Cited in ibid., 17.

Interdisciplinary dialogue, especially between traditional and disparate areas such as theology and science, is somewhat in vogue within prestigious universities and institutes.[20] Detailed and respectful dialogue between academics in the science-religion area has been ongoing for many decades now. However, this is not always appreciated by the media or in some church traditions, particularly if one's starting assumption is that science and faith are incompatible. In addition, excellent books have been written on this topic; many are academic and are therefore not widely read, or their contents are not appreciated by the majority of churchgoers. This is most likely to be the case for those who feel, in some way, threatened by the findings of science. But even this is changing in some evangelical and conservative Christian traditions.

One of the reasons why some find dialogue genuinely difficult is because of their personal commitment to their own position, whether that be science or faith, and therefore have an instinctive desire to undermine the other. Genuine dialogue, however, involves a risk, a risk that one may be changed by that encounter. The pioneering missionary Vincent Donovan saw evangelism in the same light. For Donovan:

> Evangelism is an unpredictable process of bringing the gospel to people where they are, not where you would like them to be. . . . You must have the courage to go with them to a place that neither you nor they have ever been before.[21]

This view of evangelism is challenging and goes against conventional evangelical instincts and teaching. The traditional missionary stance is as the bringer of propositional truth, with the added assumption that conversion means that you will then become like "one of us."[22] Thus we *are* prepared to go "where they are" but, in the back of our minds, we have a firm vision of where we would like them to end up! This consequently inhibits our ability to truly come alongside other people, as we have a preexisting agenda. This missionary encounter is, then, not for authentic friendship where you then journey together, but one where the bringer (superior) and receiver

---

20. See, e.g., the Center for Theology and the Natural Sciences, www.ctns.org; the Vatican Observatory, www.vaticanobservatory.va; the John Templeton Foundation, www.templeton.org; the Faraday Institute for Science and Religion, www.faraday.st-edmunds.cam.ac.uk; and the American Scientific Association (and international partners), network.asa3.org.

21. Donovan, *Christianity Rediscovered*, xii–xiii; see also 140–43.

22. Newbigin makes the same point; see Newbigin, *Open Secret*, 122–24.

(inferior) of the gospel have a very different status. Since the missionary expects the recipient to *change*, but *not* they themselves, there is no true dialogue. If, on the other hand, we embrace the *risk* of also being changed, then this makes the encounter more open-ended allowing us to go on a genuine journey to where neither of us has been before. This becomes a journey of *faith*, a faith not in the contents of the message, as such, but one that trusts the Holy Spirit to reveal the essential truths to both of us in the dialogue encounter.[23]

INTEGRATION

Extending dialogue even further, one can have a more systematic and extensive type of partnership between science and religion that Barbour called integration. The goal of integration is to explore areas where the contents of science and theology may have convergence such that doctrines can be reinterpreted in the light of the findings of science. There are obvious dangers in this endeavor, including eisegesis, where one reads the discoveries of modern science *into* Scripture. In addition, if, or when, there is a new paradigm shift in science, this will necessarily impact on those revised doctrines. As it has in the past, this can create trauma in the life of faith. However, doctrines are continually being reexamined and revised. The Reformed Church's traditional adage is "reformed and always reforming." Doctrines are sensitively recontextualized in each generation and cultural location with theology's systematizations being informed by Scripture, reason, tradition, and experience. Consequently, while dangers exist, integration can be insightful and faith-enhancing, and therefore worthy of exploration. Barbour identifies three versions of integration, namely: Natural Theology, a Theology of Nature, and a Systematic Synthesis of Theology, Science, and Philosophy.

Natural Theology

The long tradition of natural theology has sought for proof in nature, or at least for strong suggestive evidence, for the existence of God. In contrast to

23. The traditional approach is ecclesiocentric, in that becoming like "us" is inward-focused at building the church and based on "no salvation outside the church" (*extra ecclesiam nulla salus*). Donovan's and Newbigin's approaches are Christocentric (or Trinitarian), in the sense that they focus on Christ and not the institution.

a theology of nature, which will be discussed in the next section, natural theology begins with science and nature and, from *that* starting point, seeks to infer something about "God" based on human reason. (I place "God" in quotes as it will become clear that the "God" so deduced is not necessarily the God of the Bible.) In contrast to independence, natural theology protects us from an over-emphasis of interior, existential faith.[24] Polkinghorne reminds us that natural theology also counters God's utter transcendence from the created order (Barth) with a quest for signs of immanence, since creation is a potential vehicle for God's self-disclosure.[25]

Typically, natural theology starts with the classic arguments of First Cause (Cosmological) and Design (Teleological), discussed extensively by Thomas Aquinas and modernized by such theologians as Richard Swinburne. These arguments are more philosophical than theological or scientific, even though they pertain to the natural order. I have therefore summarized and briefly critiqued these traditional theistic arguments in appendix 1. This section outlines and discusses two modern versions of the argument from design, namely, the anthropic principle and intelligent design.

The Anthropic Principle

One of the most surprising things that astronomers and physicists have discovered is the level of "fine-tuning" that seems to exist in the universe. Here are *just* three examples:[26]

1. In the earliest moments after the big bang the quantities of matter and antimatter were almost in equivalent amounts. But the symmetry was not perfect; for about every billion pairs of quarks and the anti-quarks there was one extra quark. The quark-antiquarks pairs annihilated each other liberating radiation. Had there been perfect symmetry, there would be just radiation; galaxies, stars, and planets would never have come into existence. It is not unreasonable to marvel at this tiny asymmetry in the initial conditions and wonder why this was the case.

---

24. Polkinghorne, *One World*, 82.
25. Polkinghorne, *Science and Creation*, 2–3.
26. See, e.g., Barbour, *When Science Meets Religion*, 57–59; McGrath, *Science and Religion*, 181–86; Collins, *Language of God*, 71–78; Stannard, *Science and Belief*, 69–92; Peterson et al., *Reason and Religious Belief*, 106–8; Polkinghorne, *One World*, 56–58; 79–81.

2. The expansion rate of the universe itself depends on many factors, such as the initial explosion energy, the mass of the universe, and the strength of the gravitational force. Stephen Hawking writes, "If the rate of expansion one second after the big bang had been smaller by even one part in 100,000 million million it would have re-collapsed before it reached the present size."[27] On the other hand, if it had been greater by one part in a million, the universe would have expanded too rapidly and the distribution of matter would become too tenuous for the stars and planets to form. There is, therefore, an extremely sensitive balance between the explosive force of the big bang and the strength of the gravitational force. This fine balance has been estimated to be to within one part in 10^{60}! As Paul Davies says: It is the same as aiming at a target an inch wide on the other side of the observable universe (20 billion light years away) and hitting the mark![28] It is no wonder Davies says: "There is more to the world than meets the eye."[29] The expansion rate of the universe seems to be balanced on a knife edge.[30]

3. In addition, there is an extremely delicate balance between the relative strengths of the strong and weak nuclear forces. If the strong nuclear force had been slightly weaker there would only be hydrogen in the universe. On the other hand, if the strong force were slightly stronger, all the hydrogen would have been converted into helium and there would be no long-lived stars (or hydrogen) essential for the emergence of life. In addition, the weak nuclear force enables neutrons and protons to transform into each other, and to combine to form deuterium. The magnitude of this weak force, and its relative strength to the strong nuclear force, are both critical for star formation and the regulation of the rate at which hydrogen is converted into heavier elements. The fine balance between these forces determines the helium to hydrogen ratio, as well as the possibility of the exploding dying stars (super novae). These are essential features, as water (made from hydrogen) and carbon, oxygen (and other elements) are prerequisites for life. This balance is incredibly fine!

---

27. Hawking, *Brief History of Time*, 121.
28. Cited in Polkinghorne, *One World*, 57.
29. Cited in ibid., 64.
30. Barbour, *When Science Meets Religion*, 57.

The recognition that the universe is finely-tuned in terms of the balance of the forces of nature for the existence of life on earth is the so-called *Anthropic Principle*.[31] The origin of the principle's name is the obvious fact that *we* are the observers of this universe and that the cosmos itself has a Goldilocks "just right" quality for our own existence. The sheer number of coincidences has resulted in the anthropic principle being widely discussed among scientists, philosophers, and theologians. It naturally leads to a modern form of the argument from design. For example, physicist Freeman Dyson concludes: "The more I examine the universe and the details of its architecture, the more evidence I find that the universe in some sense must have known we were coming."[32] Of course most scientists would disagree. Physicist Paul Davies's nuanced caution is, perhaps, more representative: "The seemingly miraculous occurrence of numerical values that nature has assigned to her fundamental constants must remain the most compelling evidence for an element of cosmic design."[33] What is remarkable is that scientists themselves are making such comments, rather than philosophers or theologians.

How are we to explain this apparent fine-tuning? Collins lists three possible responses:[34]

1. It may be that there are an infinite number of independent universes, either occurring sequentially (i.e., an oscillating universe) or simultaneously, with different values to the physical constants that determine the relative strengths of the forces of nature. In addition, the form of the equations for those forces may be different from our own universe, along with—presumably—the number of forces themselves. This is referred to as the "multiverse hypothesis." In such a scenario, one of these many universes has, by chance, just the right balance for our existence, and we simply happened to inhabit that particular universe. One cannot help but think this hypothesis is somewhat contrived or *ad hoc*. Polkinghorne concludes:

   > Let us recognize these speculations for what they are. They are not physics but, in the strictest sense, metaphysics. There is no

---

31. For detailed explorations of the Anthropic Principle see: Barrow and Tipler, *Anthropic Cosmological Principle*; Rees, *Just Six Numbers*; and McGrath, *Fine-Tuned Universe*.
32. Barbour, *When Science Meets Religion*, 58.
33. Cited in McGrath, *Science and Religion*, 181.
34. Collins, *Language of God*, 74–75.

purely scientific reason to believe in an ensemble of universes. By construction these other worlds are unknowable by us. A possible explanation of equal intellectual respectability—and to my mind great economy and elegance—would be that this one world is the way it is because it is the creation of the will of the Creator who purposes that it should be so.[35]

2. Perhaps there is only one universe and it just happens to have all the right characteristics for intelligent life to have evolved to observe it. In that sense we are just extremely lucky. However, to say "we are just lucky" is no explanation at all.[36] Nevertheless, improbable though it is, we are here. While it is true that our existence is most unlikely, it does not mean that it is *impossible* to provide a natural explanation, or develop one in the future. But it does not follow that such an explanation would be the *best* possible one, or the most *plausible*.[37]

3. Alternatively, there is only one universe and this is it, but that the physical constants and equations are not accidental but were designed. This conclusion can form the basis of a natural theology since it is based on a thorough understanding of the intelligible, tightly-knit structure of the cosmos that science *already* discerns. Because the anthropic principle is based on what is scientifically known, rather than unknown, it is invulnerable to the charge of being a return to the "God-of-the-gaps" (which will be discussed in the next section).[38]

Since the anthropic principle is based on the physical nature of the cosmos and the relative strengths of the forces of nature, it has nothing to do with biological evolution, *per se*, other than evolution itself requires such a cosmos. Like all arguments from design, the anthropic principle is unsatisfactory as *proof* for the existence of "God" since we cannot rule out the multiverse hypothesis, or the *possibility* of a natural explanation in the case of a singular universe. The nonreligious person who does not like the—literally—astronomical odds of our universe being "just right" for intelligent life has little choice but to accept that we live in one of an uncountable number of possible universes. It should also be recognized,

---

35. Polkinghorne, *One World*, 80.

36. The meaning and role of chance in science is nuanced and complex. This is explored further in chapter 5.

37. Peterson et al., *Reason and Religious Belief*, 108.

38. Polkinghorne, *Science and Creation*, 23.

however, that the multiverse hypothesis creates no problem for the theist. From a Christian perspective, it is quite reasonable to think that the God who enabled such biological diversity on our own planet could also countenance many other possible diverse and interesting universes too. Stannard makes an interesting further observation, namely, that acceptance of the multiverse idea would be the ultimate step in placing our existence here on earth in perspective. Our quantitative understanding of the *size* of the cosmos has been steadily expanding since the time of the Greek philosophers. We now believe that the sun is just an ordinary star, one of 100 billion others belonging to the Milky Way galaxy, which in turn was just one of another 100 billion galaxies in the observable universe. We are now further faced with the possibility that we may just be a life form within one of a virtually infinite number of other universes.[39] One can then indeed marvel along with the psalmist:

> O LORD, our Sovereign, how majestic is your name in all the earth! You have set your glory above the heavens.... When I look at your heavens, the work of your fingers, the moon and the stars that you have established; *what are human beings that you are mindful of them, mortals that you care for them?*[40]

Intelligent Design

Another modern idea is that of intelligent design (ID), which is *completely* disconnected from the anthropic principle despite the fact that both purportedly point to the possibility of an intelligent designer. For a start, ID is focused on biological systems and seeks for an alternative explanation to Neo-Darwinism for irreducible complexity. The founding proponents, such as William Dembski, Michael Behe, and Phillip Johnson, have an explicit Christian worldview and consequently the idea is often dismissed in secular academia as another version of creationism. But *all* ideas should be taken seriously by the scientific community and tested accordingly.[41]

As a proposed example of intelligent design, Behe considers the motion of bacteria by their flagella, which is analogous to an outboard motor.

---

39. Stannard, *Science and Belief*, 90–91.
40. Ps 8:1, 3–4, emphasis mine.
41. See, e.g., the extensive discussion in Plantinga, *Where the Conflict Really Lies*, 225–64.

On Ways of Relating Science and Christianity

A flagellum's multiple components must all function together in a coordinated way in this "nanotechnology engineering marvel," as Collins puts it.[42] Behe's argument is that such a complex device could *never* have come into being on the basis of evolutionary processes alone.[43] One aspect of the motor may well have evolved by chance over time, but not all of them together as a unit. What is therefore required is an intelligent designer to supernaturally intervene in otherwise natural processes to bring about the formation of irreducibly complex systems. This is a modern version of Paley's argument to prove the existence of a designer, whose logic is well-known to be flawed (see appendix 1).

Intelligent design has been criticized on a number of levels. One serious concern is whether ID is genuinely science or should it better be classed as metaphysics; I am inclined to the latter view. Another serious problem is that some systems that appeared to be irreducibly complex have been demonstrated not to be so, and consequently the primary basis for ID has been seriously undermined. Collins summarizes, "ID proponents have made the mistake of confusing the unknown with the unknowable, or the unsolved with the unsolvable."[44] A further scientific critique is that ID needs to move from the cellular level to that of the genetic code. Genetics is a field in which scientific knowledge has made huge strides in recent decades and hence a more promising line of research for explaining irreducible complexity. Moreover, what kind of intelligent designer, who intervenes in developing biological systems, leaves the human body with wisdom teeth and an unnecessary appendix in our intestines?

Intelligent design can also be critiqued on theological grounds in that it is a "God-of-the-gaps" theory. Such a theory inserts divine intervention as an explanation for things that science cannot presently explain. This is always a dangerous position, because if at some future date a scientific explanation is forthcoming, it seriously undermines an individual's faith and discredits Christianity as a whole. For example, the presence of a total solar eclipse would have at one time have required a divine explanation, whereas now such events are entirely predictable. Collins concludes:

---

42. Collins, *Language of God*, 185. Similar arguments have traditionally proposed to account for the development of the eye.

43. As a generalization, microevolution (small modifications in existing species) is tolerated but macroevolution (the formation of new species) is not.

44. Ibid., 188.

> Intelligent Design fits into this discouraging tradition [i.e., that of the "God-of-the-gaps"] and faces the same ultimate demise.... The perceived gaps in evolution that ID intended to fill with God are instead being filled by advances in science. By forcing this limited, narrow view of God's role, Intelligent Design is ironically on a path towards doing considerable damage to faith.[45]

The proponents of intelligent design are wise enough not to simply equate this designer with God. Rather, the designer is simply another "god-of-the-philosophers" (see below). However, it would be fair to say that some Christian traditions have quickly, and firmly, hitched ID to their wagon and so the designer is automatically (and, in my view, uncritically) associated with the Creator God. As Collins and history (see chapter 1) warn us, this is a dangerous thing *for Christians*. The noted Christian philosopher Alvin Plantinga is not convinced that ID offers substantial or rigorous support for theism.[46] On a more pastoral note, if the Creator intervenes *regularly* within the created order in the way described, ID advocates need to explain why we do not see more overt evidence for divine action—especially in response to our prayers. The sincerity of intelligent design supporters within those Christian traditions is not in doubt. But, like creationism, ID is founded on an unnecessary premise, namely, that evolution *per se* is necessarily anti-God and hence is inevitably in conflict with the Christian faith. Our Creator, however, is much more inventive, subtle, and resourceful than we can imagine!

In concluding this section on natural theology, it should be pointed out that the "God" these arguments strive to support is merely an *instigator* or a *designer*. It is not even obvious that this "God" is an individual; it could be a committee of gods. Neither is it apparent that this "God" is still interested in our universe as a whole, or our planet in particular. The notion of a clock-maker God who made the cosmos, wound up the mechanism, and then let it proceed on its predetermined course is known as deism. Deism was very popular in the Enlightenment and many who claim to believe in God today have this kind of deity in mind. This creator "God," though powerful and intelligent, is far from the relational God of the Judeo-Christian tradition.

Natural theology's use of reason can give valuable and encouraging insights to those who *already* have faith. Yet, despite Paul's bold assertion

---

45. Ibid., 193–95.
46. Plantinga, *Where the Conflict Really Lies*, 262–64.

in Romans 1:20, natural theology is of little value to the ardent skeptic, since its conclusions lack total certainty.[47] At best we can conclude that the findings of science *may* point to a Creator and that it is not unreasonable to believe in the existence of such a "God." However, although nature displays beauty and order, it also exhibits a darker side of suffering. What is termed "natural evil" exists (e.g., disease, disability, decay, and death) and this cannot simply be glossed over in a theology derived *from* nature. Moreover, while evolution tends to emphasize steady progress, we cannot overlook the role of chaos and catastrophe on our geologically active planet (e.g., volcanoes, earthquakes, and tsunamis), together with ice ages, giant meteor collisions, and destructive weather systems. What does *this* evidence say about "God"?

Finally, I offer a brief caution in the use of natural theology in a church context. Christian apologetics uses aspects of natural theology as part of its stock in trade. Hugh Montefiore writes:

> While it is true that cold intellectual thinking can never bring anyone into a warm personal relationship with God, it is also true that, while a subjective commitment to God may be satisfying to the self, it lacks credibility to others unless it can be shown that there are good reasons for the actual existence of the God to whom commitment has been given.[48]

While I totally agree with Montefiore, I have some concerns over the potential abuse of certain styles of apologetics. First, it is all too easy to have a highly selective use of the positive features of natural theology, such as the anthropic principle, and completely overlook the issue of natural evil—or that of deism. Our use of natural theology in apologetics needs to be honest. Second, and not unrelatedly, while wise advocates of apologetics appreciate its limitations, its arguments can be misunderstood by the undiscerning ear as providing proofs for those on a quest for *certainty*. In my view such things belong in the world of modernism, not today's postmodernism, with

---

47. "Ever since the creation of the world his eternal power and divine nature, invisible though they are, have been understood and seen through the things he has made. *So they are without excuse* . . . ," Rom 1:20, (emphasis mine). Polkinghorne, commenting on this verse, writes: "We might feel that the clarity of the case is somewhat exaggerated by Paul but his words certainly encourage the attempt to pursue a natural theology. God is the elusive hidden one, not overpowering us by his unveiled presence, but it would surely be disconcerting if there were no signs of him to be found in his creation." Polkinghorne, *Science and Creation*, 7.

48. Cited in Polkinghorne, *Science and Creation*, 3.

the former's emphasis on the rule of reason. Good reasons *can* be given for our faith, but these Christian "responses" should not be misconstrued as definitive or unique "answers." I certainly do not dismiss apologetics out of hand, it is a valuable nuanced tool within evangelism and faith-building that complements the work of the Spirit, but it has been over-emphasized in some Christian traditions. Nevertheless, natural theology is a potentially fruitful endeavor that gives strong hints of a Greater Reality. From a Christian perspective, the deduced character of that Greater Reality from nature can only ever be partial. Only God's own *self*-disclosure can make Godself fully known.

Concordism

As a segue toward a formal theology of nature, it is important to comment on concordism—a form of integration that is popular in some church traditions. It is based on the premise that "all truth is God's truth" and accepts some of the findings of modern science. Proponents of concordism naturally view Scripture as God's truth and therefore seek to harmonize the two together. One example is to interpret the days of Genesis as extended periods of time (cf. geological eras) rather than literal 24 hours. This implies Scripture contains a *scientific* account of our origins, an assumption that concordism shares with creationism. Concordism endeavors to take science and the Bible seriously, and tries to make the Scriptures relevant for today's context. Positive though those features are, concordism can be critiqued in several ways.

First, concordism is *selective* in which findings of science it tries to harmonize with Scripture. Proponents tend to be more comfortable in embracing cosmic evolution than biological evolution. Since concordists themselves determine what scientific conclusions to adopt, they have set concordism up as an alternative *authority* to the scientific community. Concordism, therefore, is not born out of genuine "dialogue" with science, but approaches science with a closed agenda.

Second, concordism's interpretation of Scripture focuses *exclusively* on present sense-events, not founding sense-events.[49] It appeals to the "fuller sense" of Scripture, namely, that there are additional meanings to the text, intended by God but not by the original human author.[50] Concordism,

---

49. This terminology was introduced and discussed in chapter 2.

50. See also the discussion in chapter 2 on Barthes's "death of the author" as a critique

then, stresses God as the ultimate author of Scripture. But the danger with this approach is that *we* make the text say something that it never said.[51] In the context of science, it is as if God by-passed the biblical author and his original audience, and encodes within the creation narratives a message that can only be unlocked in the light of twenty-first-century science. What does this say about exegetes of earlier centuries, who did not have the benefit of modern discoveries? What does it say about the character of God, or the way the Spirit works?

I agree with Walton: the worldview of Genesis 1 is that of ancient cosmology in keeping with the perspectives of the time. The author of Genesis 1 is not presenting an apologetic case, arguing for his cosmic perspective over others; he is simply stating his own view.[52] "There is not a single instance in the Old Testament of God giving scientific information that transcended the understanding of the Israelite audience."[53] As mentioned earlier, neither do we find reference to Greek science in the New Testament—even something as basic as the world is spherical—rather than trilayered, with the domed heavens above, and *sheol* below. Walton concludes:

> Since God did not deem it necessary to communicate a different way of imagining the world to Israel but was content for them to retain the native ancient cosmic geography, we can conclude that it was not God's purpose to reveal the details of cosmic geography.... The shape of the earth, the nature of the sky, the locations of the sun, moon, and stars, are simply not of significance, and God could communicate what he desired regardless of one's cosmic geography.[54]

Concordism's approach intentionally attempts to interpret the ancient text in modern terms. Ironically, Bultmann's demythologization of Scripture was also a means of making biblical culture more accessible and relevant for a modern reader. I am sure concordists would decry the comparison. In post-critical hermeneutics it is perfectly legitimate to explore biblical interpretations for today's context. Those interpretations, however, are not at liberty to bend out of shape a scholarly interpretation of Scripture in its

---

of Schleiermacher's confidence in the "knowability" of the author from studying the text and its context.

51. Walton, *Lost World of Genesis*, 15.
52. Ibid., 103.
53. Ibid., 105.
54. Ibid., 16.

original context—as best as can be ascertained through historical critical methods. As I mentioned in chapter 2, an exegetical study of the founding sense-event acts as a kind of "guardrail" to present sense-events. In my mind, concordism's hermeneutic is well-meaning, but ultimately flawed.[55]

A Theology of Nature

In contrast to natural theology, a theology of nature begins with a religious tradition, in our case Christianity, and revises some of its traditional doctrines in light of the discoveries of science. Consequently, a theology of nature is a significant move beyond dialogue, as it endeavors to *integrate* science and faith. Various Christian doctrines that can be so revised include that of creation, providence (God's activity in the world), and human nature, as well as generally refine our understanding of God, and the age-old problem of evil. All of these doctrines address, in one form or another, God's relation to his creation.

A theology of nature embraces the broadly accepted findings of science, respecting them as being derived from another scholarly tradition. It does not endeavor to force or bend those findings out of shape to fit a preconceived theological perspective. Instead, theologians respect the integrity of the scientific community and simply receive their conclusions, rather than opposing or dismissing them. Integration occurs, then, through the process of respectful dialogue and by allowing *well-established* scientific discoveries to challenge traditional and cultural presuppositions of theology. The theologian, therefore, embraces the big bang, a 14-billion-year-old cosmos, and a 4.5-billion-year-old earth. The difference between those two ages infers that we are all made from the ashes of previous stars. The Christian also accepts that life has evolved through a long process of emergent novelty characterized throughout by chance and necessity.[56] From science we learn that our planet contains biological and physical processes that are interdependent and multileveled; as people of faith we admire the wonder and give glory to God.

---

55. While there are various forms of Concordism, I respectfully maintain that they generally suffer from the same flaw. It should be noted that highly respected theologians on science-faith dialogue—like John Polkinghorne and Alister McGrath—make no mention of concordism as an example of dialogue or integration. Their silence on this matter shouts loudly.

56. Barbour, *Religion and Science*, 101.

On Ways of Relating Science and Christianity

In light of the present ecological crisis, it should come as no surprise that the traditional doctrine of creation is a topic of critical discussion and revision. In an influential article published in 1967, Lynn White Jr. places the blame for the present ecological crisis firmly at the feet of Christianity.[57] Even though London had a smog problem as early as 1285 from the burning of soft coal, the present scale of damage to the environment—and our *potential* to damage it—is now virtually out of control. White writes: "No creature other than man has ever managed to foul its nest in such short order."[58] The reason for the rapid ecological demise over the last few centuries is the marriage of science and technology, which has given humankind enormous power over nature. White connects that marriage, sanctified within a Christian worldview, with the traditional exegesis of "dominion" in Genesis 1:26. The language of "power *over* nature" that "dominion" and "subdue" conjure, together with the notion that such control is divinely mandated, has shaped the historical Christian doctrines of creation and of human nature. The notion of being made in the image of God (*imago Dei*) has given humankind a kind of transcendence, resulting in a distance or separation from the rest of creation and mastery over it. While I think White's critique is overly harsh in blaming Christianity for the ecological crisis, he still makes a fair point over the issue of exegesis and doctrines.[59]

In response, the notion of *stewardship* of nature recognizes that the world ultimately belongs to God who made it. Our role is to be responsible trustees and to be accountable to God for our treatment of the created order. At the beginning of Genesis 2 we read of God's week of creative activity ending in the Sabbath. In Leviticus 25 we learn of the ideal of Jubilee. Both speak of rest and restoration for humankind *and* the created order. Since stewardship involves accountability to the One who gives us that responsibility, we should not then distort our role with a merely utilitarian view of nature. Instead, a *celebration* of nature acknowledges that the created

---

57. White, "Roots of Ecological Crisis," 1203–7.

58. Ibid., 1203.

59. His analysis, in my view, overlooks the significance of the Enlightenment that rejected traditional notions of authority, including God, and replaced it with a new aim. That quest can be summarized as establishing a self-constructed personal identity and destiny based on a rational, autonomous individual that then led to a social contract to construct a progressive society. The success of science and technology fueled that goal. In light of that, it is unreasonable to place the guilt for today's ecological crisis *solely* at Christianity's feet. The sociopolitical (and religious) context of modern science is complex and science and technology are not neutral as White seems to imply.

order has inherent value. This correlates well with the Creator's affirmation that the creation is "very good" (Gen 1:31), and we also recall that God's covenant with Noah includes all living creatures (Gen 9:8–17). Certain Christian traditions go further and view nature in a *sacramental* way, such that you find God's presence within nature. Some theologians, like Moltmann, recognizing the Spirit's role in creation (e.g., Gen 1:2; Ps 104:30), emphasize God's immanence by incorporating the Holy Spirit *into* nature.[60] Panentheism, where God includes the cosmos but also exceeds it, makes nature *sacred*. In this view, the cosmos is analogous to a fetus, distinct from the mother yet dependent on her for life. Nevertheless, we can speak of the pregnant mother and the fetus as united, since the child is not yet born. Interesting though this idea is, Polkinghorne cautions:

> The problem then lies in the danger that such a view compromises the world's freedom to be itself, which God has given to his creation, and also the otherness that he retains for himself. . . . There are distinctions between God and the world that Christian theology cannot afford to blur. They lie at the root of the religious claim that meeting with God involves a personal encounter, not just a communing with the cosmos.[61]

The traditional cry of the Old Testament prophet is for social justice and a return to true worship of God. Environmental ethics is one element of social justice and arises from a revised theology of nature, however that may be formulated, and is a timely corrective to the traditional "domination" doctrine that elevates humankind above creation. Indeed it can be argued that the desire to be above creation and more like God is one aspect of the Fall story (Gen 3), and that of the Tower of Babel (Gen 11:1–9). I will explore these creation texts further in chapter 8.

There are other Christian doctrines that are being reexamined and revised in the light of our discoveries of science, such as God's providence. Aspects of that topic, namely, God's continuing involvement with the processes of creation, and in miracles and through prayer, will be discussed in chapter 7. We can also ask questions about God's relationship with time, in the light of our revised understanding of space-time from Einstein's

---

60. See Moltmann, *Science and Wisdom*, 119–24.

61. Polkinghorne, *Science and Providence*, 20. Elsewhere he writes: "I do not accept panentheism (the idea that creation is in God, though God exceeds creation) as a theological reality for the present world, but I do believe in it as the form of eschatological destiny for the world to come" (see Polkinghorne, *God of Hope*, 114–15).

Relativity. Or, how does biological evolution inform the doctrine of human nature, i.e., being made in the image of God? Furthermore, given the presence of suffering, we can explore our understanding of God's sovereignty and omnipotence. These are all good questions for further dialogue which might result in insights that eventually become integrated into Christian doctrines.

Systematic Synthesis

The development of a theology of nature is one example of integration that can occur between theology and the physical and life sciences. A *systematic synthesis* also incorporates philosophy and gives a more general framework that includes its own comprehensive metaphysics. One relatively modern synthesis is Process Theology. Before we can discuss that perspective, features of Thomism and deism will be discussed briefly to provide context.

Thomism and Deism

Thomas Aquinas, adapting Aristotle's four causes, emphasized *primary* and *secondary* causality. God, the unmoved mover, is the primary cause and works through secondary causes. What we call the "laws of nature" are examples of those secondary causes *through* which God acts in the world. In that sense science only investigates the secondary causes, with no access to the underlying primary cause. This advocates for independence—rather than integration—between theology and science, and emphasizes the transcendence of God.

The connection between primary and secondary causes is subtle. Some have made the analogy of a carpenter with a hammer hitting a nail, or a pianist playing music on a piano. While the persons are the primary causes, the nail being driven into the wood, or the music we hear, are both due to the intermediary instruments which are the secondary causes. Yet these analogies are misleading because Thomism demands that secondary causes have genuine reality in their own right, i.e., they are autonomous. While both causes are required, each has a different role to play.

God, then, does not act *directly* in the world, but through a chain of events that God initiates. God has *delegated* causal efficacy not only to creatures but also to the natural order. The Jesuit scientist William Stoeger writes:

> If we put this in an evolutionary context . . . we can conceive of God's continuing creative action as being realized through the natural unfolding of nature's potentialities and the continuing emergence of novelty, of self-organization, of life, of mind and spirit.[62]

God's purposes are built into the potentialities of nature, but God also continues to sustain the whole system and holds it in being. Without God it would cease to exist.[63]

Another key point is that the ability of the primary cause to achieve the desired end is *totally* dependent upon the intermediate, secondary cause. If that tool is flawed or deficient—like an out of tune piano, in the earlier example—the outcome will not be what the primary cause desires. This aspect of secondary causes has great relevance to the problem of evil. Suffering and pain need not be ascribed to the direct action of God, but to the frailty of the secondary causes through which God works.[64] Whether this traditional argument is valid or coherent is a matter of debate in theodicy.

While Thomism still has many advocates, especially in the Roman Catholic Church, historically the success of science led to deism. Pierre Laplace (1749–1827), while explaining the complexities of planetary motion to Napoleon was asked why he never mentioned the Creator. Laplace famously replied, "I had no need of that hypothesis."[65] The sheer self-sufficiency of Aquinas's secondary causes led to a mechanistic—or deterministic—worldview. Events were now deemed to be *only* governed by natural causes, such as Darwin's natural selection. God's role was superfluous, other than to initiate the motion of the cosmic machine. Gaps in the scientific account were fully expected to be eventually filled by physical explanations, not by introducing God into the machine. Deism accepts the idea that a rational God created the universe. However, deism argues, that the cosmos has a self-sustaining design that requires no continual intervention, so there is literally nothing left for God to do.[66] Theologians are faced with the challenge of articulating, with clarity and coherence, God's action in the

---

62. Cited in Barbour, *When Science Meets Religion*, 102.

63. Ibid.; see also 159–61.

64. McGrath, *Science and Religion*, 104–5.

65. Cited in Barbour, *Religion and Science*, 35.

66. Polkinghorne highlights the difficult balancing act: "The God who is the ground of physical process is inescapably . . . a hidden God. This is where Christian theism is "necessarily tinged with deism." See Polkinghorne, *Science and Providence*, 45. For a more detailed and nuanced review of deism, see Fergusson, *Creation*, 63–78.

world in the light of deism—whose influence still lives on. This challenge is, I believe, an essential and multifaceted task for the church today.

Process Theology

Process theology, the second example of synthesis, builds upon the philosophical ideas of Alfred North Whitehead (1861–1947), Charles Hartshorne (1897–2000), John B. Cobb Jr., and others.[67] Reality itself is viewed in a different way. Instead of a set of physical objects, reality is regarded as a sequence of temporal events or occasions of experience. Each event connects with previous events. These "events" possess a degree of freedom to develop and be influenced by their surroundings. Some events have little freedom, such as the options available to a growing tree. In contrast, humans have a vast range of options in their self-determination. Nature's development takes place in the context of a background of order. That background organizing principle, essential for growth, is identified with God. Events (or entities) receive influences from other events and from God—the uniquely imperishable entity. Events are never coerced but are *influenced* and *persuaded*. This is how God is deemed to work in the world. God is affected and influenced by the world—hence vulnerable to it—but God maintains the process in an orderly, rule-abiding way.[68] The ultimate principle in reality is then "process," "becoming," or "creativity," rather than "being" or "substance." Reality is dynamic rather than static, with the best analogy being organic—rather than mechanistic. Becoming is more basic than being.[69]

In process theology, as the name implies, God is the *God of process* and hence nature's self-determination and the freedom for evolutionary change is embraced in a positive way. Nature explores and expresses the full potentiality given to it by God through the means of both chance and necessity (i.e., the laws of nature). Ultimately God is the source of novelty, change, and order in our incomplete world—one which is still coming into being. God is not just the instigator or designer of the order within the cosmos,

---

67. Some question the term "process theology" in that process *thought* is a metaphysical way of viewing reality. It is important to remember that theology and philosophy are two different disciplines that dialogue from their respective methodological foundations and function on different levels. Note too that process theology has various forms.

68. McGrath, *Science and Religion*, 105–6.

69. Rice, "Crucial Difference," 173.

but influences all that goes on without being the sole cause for any event. God's intimate relationship with the natural world is emphasized in process theology. Indeed, the cosmos is viewed as both necessary and an intimate part of the divine life (i.e., panentheism).

Process theology therefore rejects God as the *absolute* ruler of the universe, the traditional understanding of divine omnipotence.[70] Consequently, it discards the miraculous (i.e., supernaturalism) and the traditional doctrine of creation-out-of-nothing as too one-sided.[71] God is a persuader—not a God of compulsion. God works through chance, the laws of nature, human free will, and the freedom he has bestowed within the natural order. Because creation can only be persuaded and not coerced, then God does not always get his way and so both moral and natural evil can arise for which God cannot be held responsible. Process theology argues that God cannot force nature (including humankind's decisions) to adhere to the divine will or purpose, God can only attempt to influence the process of becoming. Therefore the entities within nature have the freedom and creativity that God has endowed and he cannot override them.[72]

To say that God "cannot" do something sounds presumptuous and arrogant for a finite creature to say about the Creator. It is! And for that reason alone it is prudent to be cautious. It is worth remembering that philosophical arguments are an exercise in reason and logic. Like natural theology, process theology is a bottom-up approach to understanding God; divine revelation and theological reflection will complement, even counter, that route to knowing and experiencing God. After all, the doctrine of the Trinity does not arise from natural and process theologies, which in turn are based on a classical construct of a monarchic God.

Charles Hartshorne, an advocate of process theism, proposed the idea of a *dipolar* God who enjoys both eternity and temporality. This means that God has both transcendence *and* imminence, of being *and* becoming. Hartshorne saw the notion of a dipolar God as a corrective to classical theism with its emphasis of a transcendent God in a static state of perfection.

---

70. While this might ring alarm bells to many Christians, one should recall that Alvin Plantinga's widely respected free will defense also softens a traditional understanding of divine omnipotence. Events in this world are determined by God *and* creatures with (libertarian) free will, and not by God alone. For God's desires to be realized therefore *requires* the cooperation of humankind. Divine omnipotence will be explored further in chapter 6.

71. *Creatio ex materia* and *creatio ex nihilo* will be discussed in chapter 8.

72. Peterson et al., *Reason and Religious Belief*, 163–65.

That view is one-sided, exalting permanence over change, necessity over contingency, self-sufficiency over relatedness. God is unchanging in purpose and character, but changing in experience and relationship.[73] God's essential nature is not variable, or dependent on our cosmos. God will always exist and be perfect in love, goodness, and wisdom. Hartshorne's dipolar God can, then, be viewed as an attempt to reconcile the "god-of-the-philosophers" with the God of the Bible. Polkinghorne comments:

> While it is true that the God of *becoming* is needed if God is to be responsive to his evolving and suffering creation, is also true that the God of *being* is needed if he is to be the guarantor of the order of creation and the ground of its hope. The modern scientific view of the universe, with its reliable underlying of law but flexible open process, offers encouragement to the search for a dipolar God who is the source of the world's lawfulness and who interacts with its process.[74]

I am very sympathetic toward this delicate, if not paradoxical, balancing act.

Even so, the God of process theology seems too passive in comparison to the biblical portrayal of the Divine. Colin Gunton says of process theology that "it has been described as a sophisticated form of animism."[75] Indeed, the process "God" seems more like a life force than the *personal* God of the Christian tradition. John K. Roth refers to the process "God" as a "God on a leash."[76] Thomas Long critiques process theology as producing a God not worthy of worship.[77] And John Polkinghorne's assessment of process theology is as insufficiently strong in its portrayal of God's action to make God the ground of reliable eschatological hope.[78] He also wonders whether "Whitehead's God could be the One who raised Jesus from the dead."[79] All these criticisms resonate strongly with me. While recognizing the theological merits of qualifying the traditional characteristics of God

73. Barbour, *Religion and Science*, 104, 294.

74. Polkinghorne, *Science and Providence*, 92.

75. Cited in Polkinghorne, *Science and Providence*, 19.

76. Roth, *Encountering Evil*, 125, In addition, John Hick describes the process "God" as a "finite God" (ibid., 129).

77. Long, *What Shall We Say?*, 75.

78. Polkinghorne, *Faith of a Physicist*, 66–67. Moreover, evolutionary processes alone (however divinely persuaded) will not, in my view, bring about the eschaton but requires a radically new creative act of God (see also Isa 65:17–25; Rev 21:1).

79. Polkinghorne, "Kenotic Creation and Divine Action," 92.

(i.e., immutability, impassibility, omnipotence, and omniscience—which we will explore in chapter 6), one can't help but wonder if process theology inadvertently hand-cuffs God and then loses the key. God *is* the God of the process, but not *only* the process.

SUMMARY AND CONCLUSION

There has, historically, been the "two books tradition": the book of nature, and Holy Scripture. It is in the latter that we learn about revealed theology, of God's self-revelation. The challenge for those who advocate integration is to bring theological coherence to these two grand works of God. If theology is to be informed by Scripture, reason, tradition, and experience, then this provides a much broader foundational basis than emphasized by process theology, insightful though it undoubtedly is. Any systematic synthesis that is limited to the rule of reason overlooks the full richness of the human experience and the life of faith. After all, the things that we *value* the most—like love and beauty—defy a purely rational characterization.

At the heart of the Christian tradition is Jesus Christ whose bodily resurrection is the miraculous data point that cannot be explained away. And not just the resurrection, but the whole of the "Christ event"; namely, the birth, life, death, resurrection, ascension of Jesus and, I would add, the giving of his Spirit. However one reformulates the Christian paradigm, it must pass through the data point of the resurrection and embrace its consequences. The individual's existential faith and the corporate life of worship in the Christian community both arise in response to God's initiatives. As followers of Jesus we are also commissioned with the practical mandate of working toward justice and *shalom*. *Shalom* means bringing healing and wholeness to broken individuals, communities, and our environment, together with the restoration of relationship with our Creator. Not surprisingly, these wider aspects of the life of faith are under-emphasized by—or beyond the horizon of—bottom-up thinking alone.

This chapter has introduced Ian Barbour's four ways of relating science and faith, namely: conflict, independence, dialogue, and integration. While respecting the sincerity and commitment of those in other Christian traditions that advocate the first two categories, my own inclination favors dialogue, together with partial integration in terms of a theology of nature. It is fitting to give Barbour the last word in this chapter:

All models are limited and partial, and none gives a complete or adequate picture of reality. The world is diverse, and differing aspects of it may be better represented by one model than another. God's relation to persons differs from God's relation to personal objects. . . . The pursuit of coherence must not lead us to neglect such differences.[80]

---

80. Barbour, *When Science Meets Religion*, 180.

Chapter 5

On Chance, Order, and Necessity

Again I saw that under the sun the race is not to the swift, nor the battle to the strong, nor bread to the wise, nor riches to the intelligent, nor favor to the skillful; but time and chance happen to them all. —Ecclesiastes 9:11

I have noticed that even those who assert that everything is predestined and that we can change nothing about it still look both ways before they cross the street. —Stephen Hawking

INTRODUCTION

Having discussed foundations and frameworks in the previous chapters, we are now in a position to look at specific issues of mutual interest to science and Christianity. Science can provide a description of phenomena, and theology can give meaning to events. Both are required to give a more complete picture. The view was expressed earlier that the Bible is God's means of communication, primarily in regards to salvation, and not as a textbook of science. We must therefore be mindful of the dangers of proof-texting and eisegesis in our reading of Scripture. This is especially important when we look to the Bible for insight to questions that it does not specifically address. The biblical authors were not attempting to address *our* issues and context, but the concerns of their immediate communities. Again, theology is informed by Scripture, reason, tradition, and experience. The biblical text is very important, but the other elements are not to be dismissed—else we deny the ongoing work of the Spirit.

In reflecting on points of possible tension and potential connection between science and faith, I have become convinced that one key issue is that of *chance*. Biological evolution is a contentious issue for some Christians. As mentioned in chapter 1, part of that perceived challenge is to the *place* of humankind at the pinnacle of God's created order. Coupled with this genuine concern is, I suggest, a deep distrust of the role of chance in history. We are inclined to view disorder as negative and order as positive. This chapter explores that bias and seeks to shed light on what we mean by "chance." Although I will consider chance in the context of biological evolution, I will focus on the indeterminacy that arises from quantum mechanics. As we will see, this topic brings its own unique and fascinating challenges—and opportunities—to science and theology. I will begin with order, and a brief clarification on the status of scientific "laws."

ON THE NATURE OF SCIENTIFIC "LAWS"

It is worth commenting on the nature of scientific "laws," as the term law generally has legal or moral connotations. There is, as mentioned earlier, a long Christian tradition of "two books" of revelation: the Bible and nature. One book contains the Ten Commandments (which is also a key part of the Jewish Torah) or God's rules governing human behavior. The other—the book of nature—is also authored by God and has its own laws. It can be said that, historically, scientists were trying to uncover these objective laws, or principles, with which God endowed nature in his original acts of creation. Consequently, they saw those principles as preexistent relations to be "discovered" or revealed. In contrast, as we have seen from chapter 3, such laws are now interpreted within a non-theistic framework as human-made conclusions, or scientific summary statements, on how the universe appears to behave.

There is a further complication. In the past, universal statements became known as scientific laws with the implication that laws of nature *cannot* be broken.[1] We now recognize that Newton's laws of motion have been "broken" by relativity and quantum mechanics, thus the *universality* of laws is undermined. Newton's laws are now regarded as having a range of

---

1. Recall that, in a *clockwork* cosmos, the smooth-running mechanism cannot be interrupted.

validity, rather than applied everywhere and at all times. They do not work at speeds close to that of light or in the quantum world of atoms, molecules and fundamental particles. Another well-known formula is Ohm's law ("$V = IR$"), which relates applied voltage (V) to the current (I) through a resistor (R) in an electrical circuit. In the early 1910s a new situation arose that "violated" Ohm's law and superconductivity was discovered. It is not that Ohm's "law" is wrong, *per se*, rather new situations arose (very low temperatures) where it did not apply. In this context, the principle of induction (discussed in chapter 3) is potentially limiting, as naively believing in its universality might stifle creativity that could lead to dramatic new discoveries.

In light of this, we can see that a scientific law is not something to *enforce*—as are society's laws by the police or courts. Rather, laws strictly *describe* what has happened and do not unambiguously *prescribe* what *must* or *will* happen in all circumstances. Furthermore, the earth does not go around the sun because Newton's laws of gravitation and motion *make* it do so, rather the earth has its motion and Newton's mathematical expressions are our way of describing how it travels. We therefore need to tone down our "absolutist" rhetoric, or perceptions, concerning scientific laws. For these reasons "law" is an antiquated term that is generally no longer used. Even Einstein's famous $E = mc^2$ relation is not classified as a law, though it is just as profound as Newton's laws. Moreover, it is not easy to distinguish between a law and a theory when the latter becomes noncontroversial within a specific scientific community. Quantum theory is foundational to the paradigm of modern physics, and the word "theory" should not be understood in a pejorative sense implying that it is less well established than a law.

From a theological perspective, the regularities discerned by science, that are summarized as the laws of nature, are—in fact—"signals of God's reliability and faithfulness made known in his creation."[2] The challenge is to formulate a *coherent* theology of providence that incorporates God's faithfulness in physical processes and addresses the possibility of the miraculous. Since miracles are integral to the life of Jesus, most notably his bodily resurrection, this important topic will be explored in the chapter 7.

---

2. Polkinghorne, *Science and Providence*, 10.

ON "CHANCE" IN CREATION

The opposite of order is disorder. The doctrine of creation-out-of-nothing (*creatio ex nihilo*) notwithstanding, God's creative acts can be described in terms of bringing order out of chaos. If the laws of nature are the customs of God, how are we to understand randomness and chance? The traditional theological understanding is that order is good and disorder is bad—even evil.[3] Yet we are also aware of the Christian claim that God works to bring good out of evil. Consequently what is deemed as evil can be used by God in ways we struggle to fathom. Can this assertion be also applied to the physical and biological worlds? How does a scientist understand chance and its role in the cosmos? Can that insight, by means of natural theology, be useful to the Christian? Can it be incorporated within a theology of nature and our understanding of providence? All good questions to explore.

Types of "Chance" Events

When we say something happened by chance, we are not identifying chance as a *cause* in itself but just a convenient expression for a variety of possible processes. Polkinghorne identifies four distinctly different scenarios that we deem to occur "by chance."[4] They are:

1. Random quantum events, which will be discussed at length below.

2. Small fluctuations that trigger instabilities. Even within the classical world of Newton, there is also indeterminacy as shown in the study of instabilities: chaos theory. Most dynamical systems do not have stability to small variations to their initial conditions.[5] For example, a tiny change in a starting parameter in a weather system would generally produce a dramatically different scenario (the so-called "butterfly effect"), and this divergence limits our predicting capabilities. The rings of Saturn contain irregular-shaped lumps of ice that all have chaotic tumbling motions with multiple collisions while constrained into a plane by the gravitation field of the oblate planet. This complex cha-

---

3. We will explore this perspective, along with the traditional doctrine of *creatio ex nihilo*, when we study various Old Testament creation texts in chapter 8.

4. Polkinghorne, *Science and Creation*, 48. The various types of chance are also discussed in Bartholomew, *God, Chance and Purpose*.

5. See also Polkinghorne, *Science and Creation*, 42–50, and Barbour, *Religion and Science*, 181–84.

otic motion has resulted in ordered patterns to the ring structure, with configurations that continue to evolve dynamically over time. Such large-scale patterns can be said to be self-organized.

3. The coincidental combination of two independent processes; an example is the giant meteor impact that is thought to have wiped out the dinosaurs millions of years ago.

4. The general way independent pieces of matter combine and interact with each other to produce a succession of configurations.

This latter category is analogous to the way you can shuffle playing cards to produce different hands of, say, Bridge. Each hand is uncorrelated with the previous one, as there are many possible combinations of cards. But once the cards are dealt the game then proceeds following the well-defined rules. The potentiality of the game is only discovered by playing it. Because of the initial combination of cards, some Bridge games can be boring, yet with another combination the game can be surprisingly fascinating. The combination of both chance and the rules of Bridge result in an unpredictable element to a player's experience. This analogy can be applied at multiple levels from biological evolution to galaxy formation. In each case the raw material of novelty provided by chance is explored by its means of the laws of nature. In the case of evolution, some combinations will prove to be fruitful and survive by a replication in a regularly behaving environment.

One of the curious things about many, if not most, of the equations in physics is that they are time reversible. In other words, you can track the behavior of a system forward in time and then backward again, returning to the exact original starting condition. The world in which we live does not behave in that way, which goes to show that these physics equations are idealized. As I write there is a hot cup of coffee in front of me and I can see wisps of water vapor rising above it. If I don't drink it soon it will go cold. We never experienced the reverse: heat from the room warming up my coffee or all the escaped water vapor recondensing in my mug. The degree of randomness increases in this scenario from its original more ordered state. This is well understood within thermodynamics. Dissipative forces make *irreversible* changes on the system, such as heat being lost from my mug to the wider environment. This results in a unique directionality in time; we can only go forward, not backward. It seems that in order to have an arrow

On Chance, Order, and Necessity

of time, the systems must be sufficiently complex to include an element of genuine randomness within it.[6] Such is the world in which we live.

The role of chance and necessity in the biological evolution may result in blind alleys. The arrow of time means that the process cannot be reversed and some more fruitful direction explored. The mere fact that we are here demonstrates that not all chance is destructive. Chance creates openness to the future which can, evidently, also be positive. This is because biological systems are not "closed" to the outside world, like my mug of initially hot coffee, but are continually receiving and generating energy. In addition to the subtle interplay between randomness and the laws of nature, the environment in which biological systems are evolving is also changing. If the changes in the complex surrounding conditions (e.g., temperature, radiation input, localized chemical composition and concentrations, etc.) have changed sufficiently, the opportunity for a certain kind of evolutionary change (e.g., a mutation) is lost. This means that in addition to chance and necessity there is *history*. And history, because of the arrow of time, is not repeatable. Consequently, evolutionary history involves unpredictability and irreversibility.[7]

Arthur Peacocke likes Karl Popper's idea that natural processes display certain tendencies—strong enough to be termed *propensities*—for increase in complexity and, hence, an increase in organization in living systems, even to the point of the emergence of consciousness and self-awareness.[8] For the theist, such propensities will obviously be interpreted as built in to the fabric of the cosmos and intended by the Creator. In this sense the dice is loaded or biased so that over time God's creation explores such emergent possibilities through the complex interplay of chance and necessity (i.e., the law-like framework that constrains the possibilities). Peacocke states:

> If all were governed by rigid law, a repetitive and uncreative order would prevail: if chance alone ruled, no forms, patterns or organizations would persist long enough for them to have any identity or real existence and the universe would never be a cosmos and susceptible to rational enquiry. It is the combination of the two which makes possible an ordered universe capable of developing

---

6. Polkinghorne, *Science and Creation*, 41.
7. Barbour, *Religion and Science*, 238.
8. Peacocke, *Theology for a Scientific Age*, 156.

within itself new modes of existence. *The interplay of chance and law is creative.*[9]

Evolution is creative and can produce good outcomes, such as healthy conscious beings and a diverse variety of plant and animal life, which reflect the organizing power of order. Nevertheless, it is also an untidy process, reflecting the disorder in the raw material, and can result in physical "evils" such as disease and genetic deformities. It seems we cannot have one without the other. Polkinghorne concludes:

> [We live in] a world of orderliness but not of clockwork regularity, of potentiality without predictability, endowed with an assurance of development but with a certain openness as to its actual form. It is inevitably a world with ragged edges, where order and disorder interlace each other and where the exploration of possibility by chance will lead not only to the evolution of systems with increasing complexity, endowed with new possibilities, but also to the evolution of systems imperfectly formed and malfunctioning.[10]

For Polkinghorne, the world "is endowed with an *assurance* of development" because *God* designed the system of law and chance.[11] Yet God only designs the general system; the specific details are not explicit expressions of God's will. Consequently, the theological problem of suffering is less acute because God did not predestine every event. Moreover, in an analogous way to God respecting human free will, God does not override the physical system. Rather God allows his ordained processes to develop freely. Such a statement may seem stark to the Christian at first reading, for what then are we to make of the role of prayer and the possibility of the miraculous? This issue will be addressed further in the chapter 7, but before this "free process defense" is rejected too hastily, we should reflect on the fact that God's apparent lack of "interference" within our world system *does* seem to correlate with our everyday experience. For a Christian, as opposed to a deist, the question why a *loving* God does not intervene more in addressing natural and moral evil is a persistent, troubling theological problem. One way to address this issue theologically is to explore the notion of God's "self-limitation"—or *kenosis*—because that is the logic of love.[12] For the

---

9. Ibid., 65, his emphasis.

10. Polkinghorne, *Science and Creation*, 48–49.

11. Peacocke makes exactly the same point: "*God is the ultimate ground source for both law ('necessity') and 'chance.'*" Peacocke, *Theology for a Scientific Age*, 119, his emphasis.

12. See Polkinghorne, *Work of Love*, and Oord, *Uncontrolling Love of God*.

moment, the free process idea correlates well with the God who originated the cosmos and sustains continuous creation. But that does not mean it is the last word on divine action in the world.

Indeterminacy and the Quantum World

Laplace had a grand vision of a super-computer which, given all the positions and momenta of all the particles in the universe, could apply Newton's laws of motion and so discover the destiny (and derive the past) of the universe. Consequently, the scientist would at that instant *know* all the future (and history) of everything. But his dream can never be realized for several reasons—not least because the computer would need to be larger than the entire universe just to store the parameters, let alone run the program! More significantly, nature is not as deterministic—or mechanistic—as Laplace imagined.

In Laplace's worldview, matter was made up of particles—atoms—in motion. One of the obvious properties of a particle is that it is localized in a small region of space. This is to be contrasted with wave motion, which is distributed over a wide region of space (i.e., delocalized); light was widely understood in terms of wavelike behavior. In the early part of the last century, new experiments revealed situations where particles, like electrons, behave like waves, and light behaved like particles. From a phenomenological perspective, a particle is completely incompatible with a wave—as we have just seen in terms of localization. These experimental observations led to fundamental questions about the nature of matter: is it particle-like or is it wave-like? The "answer" is not straightforward. If you ask a particle-like question (i.e., perform an appropriate experiment using particle detectors), you get a particle-like answer; if you ask a wave-like question you get a wave-like answer. This paradox led to the notion of "wave-particle duality." It was not long before quantum mechanics was formalized mathematically and the motion of "electrons" around the nucleus of an atom was articulated in terms of a *wave* equation.[13] The earlier image of an atom was analogous to a mini-solar system, with electrons orbiting a heavy, central nucleus. But this representation is very misleading; the atom of the quantum world cannot be pictured at all. At best, the patterns of the waveforms—the set of mathematical solutions to the atom's wave equation—can be interpreted

---

13. Because of our preconceived ideas of particles and waves, there is great linguistic difficulty in speaking of atoms, let alone mentally visualizing them from our descriptions.

in a *probabilistic* way. The three-dimensional probability distributions are analogous to the visualizations of complex musical tones or optical intensity distributions; but these comparisons are inherently inadequate. This is because atoms are inaccessible to direct observation and unimaginable in terms of our sensory qualities. It cannot even be described coherently in terms of classical concepts such as space, time, and causality. The behavior of the very small is radically different from that of everyday objects.[14]

Quantum mechanics has, at its heart, the famous Heisenberg uncertainty principle. This states that the position and momentum of a quantum particle, like an electron, cannot be simultaneously known.[15] This is in stark contrast to Newtonian physics, on which Laplace's worldview was based, which assumed that both quantities—position and momentum—were *always* knowable. In addition, we all know today that unstable nuclei, like uranium, are subject to radioactive decay. The accepted view of the majority of physicists is that there is no assignable cause for the decay of a specific radioactive nucleus. These effects, however arbitrary on the microscopic scale, do not imply a total lack of predictability in macroscopic systems. In a large ensemble we can reliably predict that after a certain time, known as the half-life, 50 percent of the nuclei will have decayed. This is analogous to mortality statistics; a life insurance company does not know which individual might die over a ten-year period, but they are able to predict the outcome for a large group of people—and make a financial profit! There is a difference, however, because there *are* causes for an individual's death, even if they are not known to the actuary. However it is asserted that there are no causes for individual events in the quantum world.[16] How are we to interpret this apparent indeterminacy in nature?

One approach is to say that, like the actuary, the uncertainties reflect our lack of knowledge about the system. Consequently, the Heisenberg uncertainty principle is interpreted as a statement of present *ignorance*. The underlying assumption is that exact, deterministic laws do exist, it is just that we have yet to discover, or formulate, them. Einstein was someone who held this view and famously said, "God does not play dice," to which Bohr replied, "Stop telling God what to do"! Taking a different approach, Bohm postulated additional "hidden variables," as yet undetected, in an attempt to preserve the deterministic world of classical physics. His calculations

---

14. Barbour, *Religion and Science*, 167.
15. There is a corresponding uncertainty principle between energy and time.
16. Polkinghorne, *One World*, 10.

On Chance, Order, and Necessity

have yet to produce conclusions that differ from those of quantum mechanics. Most scientists are dubious about such mathematical proposals and see them as contrived. There is already a sense of elegance and economy in the equations that describe the quantum world; why spoil it? Oddly, perhaps, mathematical simplicity and beauty has an appeal to the physicist.[17]

An alternative suggestion is that Heisenberg's uncertainty principle is not about our ignorance, but an inherent, fundamental limit to experimental capabilities. Bohr and Heisenberg formulated good arguments that demonstrate that the very act of measuring the particle's position inevitably perturbs its momentum, and vice versa. (This is something that can be avoided in the everyday world of classical physics.) Consequently, the problem is linked to the process of observation. This explanation does not apply in the case of spontaneous nuclear decay, but even this can be placed under the umbrella of the uncertainty principle as representing a fundamental limit to *attainable* knowledge. Another suggestion is that we are facing a conceptual limitation. In trying to describe matter as a particle or a wave we are using language and concepts that are inherently inappropriate because they belong to the paradigm of classical physics. Because we are using the wrong linguistic analogies we must therefore be agnostic concerning any conclusions that pertain to causation and indeterminacy.

The final perspective is the one that many, if not most, physicists adopt. Namely, that Heisenberg's uncertainty principle reveals an indeterminacy that is *inherent* within nature, rather than a statement of our ignorance, or pertaining to experimental or conceptual limitations. This is simply the way the world is. (See appendix 2 for further discussion from a philosophical perspective.) Instead of the traditional quest for rigorous certainty, we must therefore embrace the *probabilistic* or *statistical* aspect to nature that arises from the quantum world. Barbour writes:

> [Experimental] observation consists in extracting from the existing probability distribution one of the many *possibilities* it contains. The influence of the observer, in this view, does not consist in disturbing a previously precise though unknown value, but in forcing one of the many existing potentialities to be actualized.

---

17. One can legitimately explore what this observation implies, for it is not *necessary* that the equations should possess such subtle properties. There is more to this issue than meets the eye. It could be said that the Copernican system was simply an alternative perspective to that of Ptolemy, since it was just as complicated—mathematically. How, at the time, were they to assess which model was "right"? (See appendix 2 for further discussion on instrumentalism and realism.)

The observer's activity becomes part of the history of the atomic event.[18]

In exploring the consequences of indeterminacy, it is important to note that since the quantum description of the present state of matter is as a range of *possibilities*, then *the future is undecided*. Barbour continues:

> More than one alternative is open and there is some opportunity for unpredictable novelty. Time involves a unique historicity and unrepeatability; the world will not repeat its course if it were restored to a former state, for at each point a different event from among the potentialities might be actualized. Potentiality and chance are objective and not merely subjective phenomena.[19]

A more dubious, if not bizarre, suggestion is that of Hugh Everett's *many-universes* interpretation of quantum mechanics, which is *not* to be confused with the multiverse proposal from cosmology mentioned in chapter 4. Every time the quantum system has more than one possible outcome, the universe splits into many separate universes, in each of which one of the possible outcomes occurs. We happen to be in the universe in which there occurs the outcome that we observe, and we have no access to the other universes in which duplicates of us observe other possibilities. Given that the vast number of quantum events in the cosmos, the universe would have to be continually dividing, resulting in an unbelievably mind-boggling large proliferation of universes. And how can we test this fanciful proposal? We can't—it is unfalsifiable. I agree with Barbour: "It seems much simpler to assume that the potentialities not actualized in our universe are not actualized anywhere."[20]

All but the first interpretation of the Heisenberg uncertainty principle reject the strictly clockwork view of the universe, but for different reasons. If the last description is correct—and indeterminacy and chance are *inherent* features of nature—then this has significant implications for theologies of nature and challenges the traditional understanding of God's omniscience. If the future is *open*, then what does this say—if anything—about God's relation to time? Is the future then open for God too? What about prophecy, purpose, and predestination? All good questions to ponder that may lead to

---

18. Barbour, *Religion and Science*, 172–73.
19. Ibid., 173.
20. Ibid.

conflict, dialogue or, potentially, integration. We will explore God's relation to time further in the next chapter.

Indeterminacy and Theology

In the world conceived by Newton and Laplace, nature was an intricate and harmonious machine that followed unchangeable laws. Those laws can be understood theologically as expressing the faithfulness of God and demonstrating his sovereignty. This is quite consistent with Aquinas's primary cause, with the laws of nature being God's instruments to achieve predetermined purposes.[21] Within the paradigm of classical physics it is quite straightforward to find coherence with the theological doctrine in predestination and the traditional view of a God who foresees everything—a God who is in total control. Just as Aristotelianism had been absorbed within the Christian worldview and so contributed to the conflict between Galileo and the church, so we can ask: has the paradigm of classical physics been uncritically absorbed into our theology? If questions concerning God's sovereignty, immutability, omniscience, and omnipotence seem destabilizing, is it because—at least in part—the *certainty* that is inherently implied within a mechanistic worldview has crept into our theological thinking and biases? These classical attributes of God will be examined in the next chapter, but—for the present—they can be seen to resonate with a deterministic worldview. If the paradigm of classical physics has influenced our view of God—and clearly the rise of deism demonstrates that it did—then it is right and proper to explore the challenge(s) of the new paradigm of modern physics to theology. The case of Galileo proved, ultimately, to be a corrective to the church's outlook—not least in terms of hermeneutics. If we believe that God is at work in history, as I do, then we have grounds to expect the present science-theology interface to be a similar enlightening work of the Spirit.

If beneath the apparent regularity of nature there lays indeterminacy within the quantum world, then *both* order and chance are manifest in creation. As we have noted, in popular usage, chance generally implies the opposite of intelligence: random, chaos, disorder, accident, purposelessness. In itself, chance is *random*, whereas divine action is said to be purposeful and goal directed.[22] Bertrand Russell wrote:

21. Barbour, *When Science Meets Religion*, 71.
22. Ibid., 72.

> Man is a product of causes which had no prevision of the end they were achieving; his origin, his growth, his hopes and fears, his loves and his beliefs, are but the outcome of accidental collocations of atoms.[23]

And Nobel Prize–winning biologist Jacques Monad wrote:

> Man knows that he is alone in the universe's immensity, out of which he emerged only by chance. . . . Chance is the source of all novelty, all creation.[24]

Both of these positions are materialistic and reductionist in outlook; we are but material machines. While conflict with those philosophical perspectives is inevitable for the Christian, there are more positive ways of viewing chance within a theological framework.

One possible theological response is that God brings about the kind of cosmos he wants out of the range of possibilities that are inherently present in the quantum world. We have mentioned before that the act of observation brings about an actuality out of the range of potentialities.[25] The infinite God is one who then brings about *his* choice out of the myriad of possible quantum states. God then is the perpetual "observer" who determines the indeterminacies. No energy or force is required, and no scientific laws are "violated" as this is, in principle, undetectable. This implies that the uncertainty principle is a statement of our ignorance, and that God is the "hidden variable" who maintains a deterministic cosmos. This view, proposed by physicist-priest William Pollard, is potentially attractive to those who are committed to predestination and to a God who is in "tight" control of the cosmos.[26] God's mode of action here is, however, *always* hidden and therefore cannot be used as an argument for natural theology—but it can be used within a theology of nature.[27]

This is an ingenious proposal that should not be dismissed lightly. What appears to be chance, which some atheists take as an argument against

---

23. Cited in ibid.

24. Cited in ibid., 73.

25. In technical language, the overall wavefunction of the system in question, which consists of a superposition of possible states (all expressed as waves), "collapses" into one outcome in the act of measurement (i.e., observation).

26. This idea is extended to include all the apparently chance events (e.g., mutations) in evolution as being predestined by God. See Barbour, *Religion and Science*, 239.

27. Barbour, *When Science Meets Religion*, 171.

On Chance, Order, and Necessity

theism, may be the very point at which God acts.[28] Nevertheless, Barbour—rightly in my view—criticizes Pollard's position in three ways.[29] First, how do we defend human free will and explain the reality of evil if God is *totally* in control, as Pollard maintains? Theological determinism ultimately, if not inadvertently, makes God the author of evil and this is, in my view, an insurmountable obstacle for this form of theodicy. Second, predestination seems to be achieved more by God's micro-control of quantum events than the orderly "customs of God" manifest in the laws of nature. While this counteracts the tendency toward deism of the latter, perhaps the pendulum has swung too far in the opposite direction. I also question whether God's "management" of all quantum events is *sufficient* for God to control the outcome at a macroscopic level and so have a completely deterministic cosmos. Third, is God restricted to bringing about his will only by means of this "bottom-up" approach? Can God not invoke "top-down" causation or input pure information by some other non-energetic means? Even at this stage, it seems prudent to allow for those possibilities, even if we have not articulated how this might arise.[30]

A further qualification to Pollard's proposal is to say that God influences *some*, rather than *all*, of the quantum events. The advantage of this is to eliminate theological determinism which undermines humankind's free will and addresses the issue of natural evil. This qualification creates its own theological problem, however, in that we are now left searching for a coherent understanding of God's action in "tweaking" (i.e., observing) some quantum events rather than others. This issue will be discussed further in chapter 7 when considering God's providence in miracles and prayer.

Yet there is something niggling about Pollard's idea. It is as if we have found a hidden back-door that allows God to enter and act in the cosmos from which he is otherwise barred by his own laws of nature! The joy of this "discovery" may appeal to those seeking a "natural" route for "supernatural" intervention. But we need to reflect on whether this is theologically credible and in keeping with the revealed character of God. Polkinghorne is also uneasy about Pollard's idea as being the principal account of God's

---

28. Barbour, *Religion and Science*, 312.

29. Ibid., 188.

30. A possible "top-down" causation would be the introduction of boundary conditions to constrain processes within certain limits. In the case of humankind, dreams and visions could be viewed as a means of imparting divine information with minimal energy input.

action in the world—it seems too contrived. He concludes: "God's relationship with his physical world must, it seems to me, be subtler than that."[31]

In contrast to Monad and Russell, and like Pollard, Donald Mackay regards chance as a *neutral* term meaning "the *absence of knowledge of causal connections* between events."[32] In other words, what we call chance is due to present human *ignorance* but the specific causes for certain events are nevertheless known to an omniscient—all-knowing—God.

> Even the events we classify technically as "chance" or "random" are *determined* by the sovereign Giver of their being. For biblical theism, nothing in the technical scientific idea of chance implies or requires any release of events from the sovereignty of the Creator.[33]

To support his view—likely held by many Christians—Mackay quotes Proverbs 16:33: "The lot is cast into the lap, but the decision is the LORD's alone," and concludes:[34]

> Could there be a clearer indication that God is the Lord of events which in this sense "happened by chance," just as much as those that seem orderly to us? It may indeed be easier for us to see God's hand in the obvious orderly pattern; but the Bible at least will not tolerate the idea that he *must* always work in this way. The "either-or" [dilemma] . . . , God *or* chance, is simply not the way the Bible relates the two.[35]

John Polkinghorne, while recognizing Mackay's (and Pollard's) view of total divine control as "logically invincible," is not wholly convinced.[36] He wonders: "Why has God chosen to hide his hand under the appearance of randomness?"[37] Polkinghorne finds the attitude of Arthur Peacocke more attractive; he sees a positive exploratory role for chance as part of the Cre-

---

31. Polkinghorne, *One World*, 72.

32. Mackay, *Clockwork Image*, 48. See also Mackay, *Science, Chance and Providence*, 25–31.

33. Mackay, *Science, Chance and Providence*, 30–31, emphasis mine.

34. One could equally well cite: Ps 33:11; 139:1–18; Isa 40:28; and Heb 4:13, to name but a few references (avoiding Eccl 9:11). This is "proof-texting" and theology is more than that! This issue also relates back to hermeneutics in chapter 2, as well as the discussion on the inspiration of Scripture. How one understands God's relation to time is also a key factor, which will be discussed in the next chapter.

35. Mackay, *Clockwork Image*, 49.

36. Polkinghorne, *One World*, 68.

37. Ibid., 68.

ator's plan to "unfold the potentialities of the universe which he himself has given it."[38] This theological response embraces *both* law (orderliness) and chance (randomness)—and their interplay—as part of God's design and providence. This "sacredizes" chance and gives randomness a positive role in the exploration of potentialities. As such, God's action is less "hands-on," less controlling, or micromanaging of creation. This perspective counters traditional theology with its emphasis on *design* and *order* as the signs of God's providential and foreordained plan, and where chance is seen as the antithesis of design. In contrast, Barbour, embracing the role of chance, observes:

> Evolution suggests another understanding of design in which there are general directions but no detailed plan. There could be a long-range strategy combined with short-term opportunism arising from feedback and adjustment. In this strategy, order grows by the use of chaos rather than by its elimination. There is improvement but not perfection. There is increasing order and information but no predictable final state.[39]

This being the case, we should heed Robert Russell's urging not to equate disorder with evil, or order with good, for disorder is sometimes the precondition for the emergence of new forms of order.[40] Rather both regularity and chance are ordained by God. Polkinghorne concludes:

> A tightly deterministic universe, evolving along predetermined lines, seems to leave little room for freedom and responsibility.... The actual balance between chance and necessity, contingency and potentiality, which we perceive seems to me to be consistent with the will of a patient and subtle Creator, content to achieve his purposes through the unfolding of process and accepting thereby a measure of the vulnerability and precariousness which always characterize the gift of freedom by love.[41]

I believe this nuanced view is a significant improvement over theological determinism. But, as mentioned earlier, this free process defense has a potential drawback if it implies that God's role is limited to originating and sustaining the cosmos. Process theism, as we saw in chapter 4, speaks of God influencing events through persuasion rather than coercion. How

---

38. Ibid., 54. See also Peacocke, *Theology for a Scientific Age*, 119.
39. Barbour, *Religion and Science*, 238.
40. Ibid.
41. Polkinghorne *One World*, 69.

precisely God "influences" events, with no applied energy or force, is not clear; but it seems as if Pollard's approach gives one possible route for bottom-up "persuasion."

In conclusion, God's action in the world has been traditionally viewed in terms of sustaining order through the laws of nature. More recently scientist-theologians are recognizing God's providential care of the cosmos through chance as well as order—i.e., through *both* contingency and necessity. Rather than viewing indeterminacy within quantum mechanics and the role of randomness in nature as a source of conflict with the Christian faith, alternative perspectives have been outlined that show fruitful potential for integration, built on the foundation of respectful dialogue. As is befitting a world that possesses a degree of openness, not every theological aspect of this picture can be tightly controlled or pinned down. I advocate that the quest for modernism's certainty, which is embodied in physical and theological determinism, needs to be abandoned. This discussion has demonstrated that God's providential action is less rigid and more fluid than has been traditionally asserted, not least by the doctrine of predestination. Living with the inherent uncertainty that this new fluidity demands is, I suggest, a normal part of our postmodern journey of faith. As mentioned before, the opposite of faith is not doubt, but certainty. We walk by faith, not by sight (2 Cor 5:7).

Chapter 6

On the Nature of God

Christian theology begins, continues, and ends with the inexhaustible mystery of God. —Daniel Migliore[1]

INTRODUCTION

In chapter 2 we explored the issue of scripture, acknowledging that its inspiration means different things to different people. It is now necessary to step back for a moment and also recognize that even the nature of God is not as straight forward as Christians might think. We have already seen, for example, that process theology's "God" is not the same as the God traditionally portrayed by the church. For the Christian, God is not derived from study of nature or philosophy but based on the biblical authors' understanding of God and his actions in history. Among other things, those writers were implicitly describing, in their own terms and contexts, what God is like and how God relates to humankind. Central to the Christian doctrine of God, then, is the scriptural witness of God's covenantal activity, first with the patriarchs and the nation of Israel, and later with the whole of humankind through the life, death, and resurrection of Jesus, the Messiah.

Since theological reflection is also informed by *reason*, dialogue with science and philosophy has an important contribution to make. Nevertheless, Christian *tradition* affirms a Trinitarian God at its very core. Yet in the

1. Migliore, *Faith Seeking Understanding*, 64.

dialogue between science and Christianity, as the astute reader will realize, the Trinity barely gets mentioned. Instead "God" is a monad, or singular, and who is perhaps best correlated with YHWH, the Creator God. Such a monarchic God was readily syncretized with Plato's "god" (*demiurge*) in the Hellenization of both Judaism *and* Christianity. Much later in history, as science gained in stature, God as *Creator* became more cerebral—moving further away from the personal God of the Judeo-Christian tradition. As we saw earlier with deism, God became regarded as the Instigator of the universe and had no further direct involvement in creation. Around the same time, Jesus became an inspirational moral teacher and the mysterious Spirit was quickly lost in the ether! Consequently, Trinitarian thinking became a thin veneer on top of the classical notion of the divine. Once this was recognized, notably by Karl Barth and Karl Rahner in the last century, a resurgence of Trinitarian thought began to occur. While I personally embrace Trinitarian theology, nevertheless, a philosophical view of God continues to strongly influence theology and its rational discourse. In this chapter the nature of this "God" will be explored since God's ascribed qualities have also been subject to challenge and revision. This is important to consider because what we assume about divine attributes not only frames the dialogue between science and Christianity, it also shapes an individual's faith in God.

A theological understanding of God will quickly move to God's transcendence and immanence. God's transcendence is his mode of being that is beyond the created cosmos. "God's being and power surpass the world and are never identical with, or confined to, or exhausted in the world that God has freely created and to which God freely relates."[2] In contrast, God's immanence is God's intimacy and closeness to all creatures, even an indwelling of all created beings.[3] We will see that there is some tension between these two poles. Traditional theism stresses God's otherness—his transcendence. This can create a sense of distance between God and creation, even a distinct separation between God the Father and the incarnate Son.

That sense of detachment gets more acute in the light of science for two reasons. First, the sheer size of the universe is beyond our imagination. If, as literal thinkers, we imagine God to be *outside* of the universe then that makes God ultra-remote—and getting more distant all the time, since our

2. Ibid., 426–27.
3. Ibid., 413.

universe is expanding. The second reason, as mentioned in the last chapter, is that we have absorbed a mechanistic image of the universe into our worldview—including our theology. This leads to viewing the cosmos as a *closed* system of pure cause-and-effect, which in turn provided the basis for deism. This "intelligent designer" is far from the relational God of Abraham, Isaac and Jacob, or the Trinitarian God of the Christian tradition. Moreover, we saw in the last chapter that the universe is more *open* than the classical physics paradigm suggests. Even so, the legacy of a closed cosmos lives on. And this further distances God from his creation such that, if he is to act within the world at all, we use the language of divine "intervention." For the modern mind, then, an emphasis on God's transcendence can lead to a God who is deemed to be outside of the universe—one that is both *closed* and *expanding*. From this perspective, it is no wonder God can seem both silent and distant.

The notion that God is outside of space and time has a long history. We consider here three classic assertions about the nature of God, namely, that God does not change (immutable), is all-powerful (omnipotent), and is all-knowing (omniscient). The first of these assertions arises *via negativa*, by considering what God is *not*—with respect to creation. And the latter depends critically on how one understands God's relation to *time*, which will be the main focus of this chapter.

GOD AND CHANGE: IMMUTABILITY AND IMPASSIBILITY

Creation is subject to *change*; everything is in a state of flux. There is natural movement; fire goes up, heavy objects fall down. As part of the water cycle, rivers flow—causing erosion—and carry silt to the sea. In addition to the annual seasons, creatures are born, grow, age, and die. And God, it was deemed, must be other than that. God is perfect and incorruptible. And perfection implies a *static* state where no change is possible, for change would either be for the worse or imply that God was lacking in some quality. Perfection, then, implies God is beyond the *realm* of change and decay—hence immutable.[4]

Many question the traditional doctrine of divine immutability on the grounds that the biblical God is a *person*. And persons are free, thinking

---

4. It is but a small step to then regard the material world as intrinsically evil, which Christianity emphatically rejects.

agents who can plan, choose, enact, and even learn. In the doctrine of *creatio ex nihilo*, God was completely free to choose to create and at liberty to decide the form that creation took. Moreover the Bible claims that God's prime characteristic is love (1 John 4:8), and love implies a give-and-take relationship whereby the lover can be affected by the beloved. This not only results in the possibility of joy and compassion, but also with the potential for suffering and grief—as love can be rejected. To say that God can suffer implies that he is a vulnerable and that what he desires can be thwarted. Many are troubled with the idea that a relational God can experience emotion and voluntarily takes such risks for the sake of love. God, they claim, is above all such sentiments and is impassive—the unmoved Mover. Others find God's intimacy comforting and endearing—and worthy of worship. In short, an immutable and impassive God is in a static state of *being*, whereas a relational God has more emphasis on divine *becoming* (e.g., God became the Creator; the Word became flesh, God will be all in all).

But, you may ask, doesn't the Bible say that God is unchanging? Yes it does (see Mal 3:6; Jas 1:17). But this does not refer to God's being, but the constancy of his character and promises. God does not change out of a deficiency of *being*, but as a result of loving relationships—change that flows from genuine interactions. From a Trinitarian perspective, God is first and foremost in the category of mutually indwelling persons, or "being-in-relationship," not that of an impersonal entity. And persons—divine or human—change in the course of relationship.[5]

GOD AND POWER: OMNIPOTENCE

The second classic divine attribute is that God is "all-powerful." But what do we mean by "omnipotent" or "Almighty," as the historical creeds describe God the Father? Fully aware of our own limitations, our instinctive response is to say: "God can do *anything*"! But is that really the case? Can God draw square circles? Can 2 + 2 = 5 for God? Can God make a stone that is too heavy for him to lift? Can God cease to exist? Can God tell a lie? Can God get married? Can God sin? This is not a linguistic game, but simply underlines the point that there are some things even "Almighty" God cannot do. Instead, a philosopher might say that God as omnipotent means he can do anything that *for him* is *logically* possible, anything which

---

5. Plantinga et al., *Introduction to Christian Theology*, 105.

On the Nature of God

does not involve a contradiction.[6] God is therefore subject to logic, just as we are, and cannot do what is logically impossible. If our understanding of God is to be coherent, this means that there must be *logical* consequences of God's decisions which impact on God himself.

The issue of logical consistency aside, many would still consider "Almighty God" to mean that he can otherwise do anything he so desires. Except, I can't help but wonder if what we mean by "all-powerful" arises from the super-human model of the divine, i.e., as a consequence of making God in *our* own image writ large. We imagine God's power to be like that of a conquering medieval monarch, only raised to the ultimate degree. This is because we have a fascination with power and of "being in control." This arises from our demand for personal rights, and a sense of self-determinism for our individual wellbeing and destiny. More collectively, we are beguiled by our sense of Western political, economic, and military dominance, and further enthralled by science and technology and the power they have provided. But is this the right approach to thinking about *God's* power? Or is this an example of a wish-fulfillment, a projection of our own desires onto the divine nature?

William of Ockham (ca. 1285–1347) described God's power in two ways. The first is omnipotence in the absolute sense we instinctively assume. God had a free choice when he considered whether or not to create the universe. Moreover, when he was considering all the possible options and variations in the kind of creation he could make, he had absolute power to realize each of them—as long as they were logically coherent. Nothing *external* was being imposed on God, but once God sovereignly decided to make *this kind* of cosmos it meant that certain other options *for this cosmos* were no longer possible, even for God. So God initially had absolute power to choose to create and what kind of cosmos it would be.[7] But, for Ockham, God cannot *now* do everything because his previous decisions inevitably limit future possibilities.[8] In other words, God's actions limit God's options. So if God made humankind with genuine (or "libertarian," to use the technical term) free will then he cannot exercise total control over us. Neither can God coerce us to freely love him. As before, we can-

---

6. Based on Richard Swinburne's definition, as cited in Hasker, *Triumph of God Over Evil*, 44.

7. We have no way of knowing if God choose to actualize any of the other possibilities!

8. See McGrath, *Christian Theology*, 210–11.

not escape the logic being expressed here. And coherence is, in my mind, important for faith. Since God has bestowed creation with a high degree of freedom, then God must be a *power-sharing* deity. Moreover, we must remember that sovereignty is not synonymous with absolute power, but rather with the total freedom God has to use his power as he wishes and for him to accomplish his purposes in the ways he sees fit.[9] Clark Pinnock writes:

> Despite having the power to control everything, God voluntarily limits the exercise of that power. . . . Almighty could mean all-determining control or it could mean a power that does not monopolize but delegates power. And, given the reality of evil in the world, God's delegation of power seems completely undeniable.[10]

This being the case, how should we understand "omnipotence"? Pinnock writes:

> God's almightiness is not an abstract domineering power. It is essentially the *power of love*. A God of love cannot be conceived in a deterministic way, like the power of the puppeteer. . . . God's power means that he is omnicompetent and can deal wisely with any circumstance that arises, not that he causes everything. It means that nothing can ultimately defeat him. . . . It takes omnipotence to create and manage freedom.[11]

Daniel Migliore—following Karl Barth—also speaks of the power of the Trinitarian God as omnipotent *love*.

> Christ crucified is the *power of God* unto salvation (1 Cor 1:23–24). The love of God made known supreme in the cross of Christ is all the power necessary to accomplish the divine purpose of creating and redeeming the world and bringing it to its appointed goal. Because God's omnipotent love is God's own, it does not work by domination or coercion but is sovereign and effective without displacing or bludgeoning God's creatures.[12]

This potency of divine love is very different from the kind of brute force we normally associate with omnipotence!

---

9. See also Walls and Dongell, *Why I Am Not a Calvinist*, 145, and Pinnock, *Most Moved Mover*, 92–96.

10. Ibid., 95–96.

11. Ibid., 94–95, emphasis mine.

12. Migliore, *Faith Seeking Understanding*, 86, emphasis mine.

On the Nature of God

GOD, TIME, AND OMNISCIENCE

The most straight forward way to conceive of God as immutable—in a static state of perfection—is to place him *outside* of time. But what does this mean? Since creation is inherently within and bound by time, we cannot conceive of what existence outside of time might entail. One possibility is that all times—past, present, and future—occur simultaneously, so to speak, analogous to God being everywhere at once (i.e., omnipresent). This means that *for God* creation was simultaneous with the exodus, the exile, and with Christ's incarnation and second coming. God simply *is*. One major problem with *any* atemporal view is that we struggle to make sense of God's relationship with creation that we read of in the biblical narrative. Within scripture it is as if God is interacting with the Old Testament heroes, talking with them and responding to them. It reads as if God is having an authentic relationship with them—at least that's how the biblical authors portray it. But how can there be any real covenantal interaction between God and the world if all of history is experienced by God at once? It has been pointed out that we would need to postulate "a two-tiered understanding of God—God as God relates to us in time, and God in God's own essential being experiencing everything at once. The latter seems to nullify and defeat the former."[13]

Of course, we can never really know how God functions with respect to time. Nevertheless, it is imperative to have some kind of tentative model in order to make our faith a living and meaningful one, and to enable meaningful dialogue with other academic disciplines. Some contemporary philosophical theologians have proposed alternative understandings of God's relation to time. For example, Nicholas Wolterstorff regards the divine life as an unending temporal sequence of past, present, and future that is separate from, but linked with, creation and referred to as life *everlasting*.[14] While some see this as radical, others regard this idea as "really quite modest":

> It holds that God is from everlasting to everlasting (Ps 90:2)—God has always been, God presently is, and God will always be (see Rev 1:4). At some temporal point in the divine life God created a world with its own history, and a different point in the divine life the Son became incarnate. In this view, since the time sequence in

---

13. Plantinga et al., *Introduction to Christian Theology*, 106–7.
14. Wolterstorff, "Unqualified Divine Temporality," 187–238.

creation and the divine life are basically analogous, it is easier to see how there could be a real history of relationship between God and creation.[15]

It is therefore quite plausible to postulate that God—at least since creation—experiences some form of temporal sequence; that he also sees succession and relates one event after another.[16]

If this is the case, what is the consequence for God's knowledge? What does it mean, then, when we say that God is omniscient? We usually say it means that "God knows everything" because we instinctively envisage God as being outside space-time and "seeing" all history at once. In this timeless scenario, God cannot help but know the cosmic past, present, *and* future.

Before considering God's knowledge in the context of his everlastingness, we need to be a bit more precise in our definition of omniscience and define "everything." A more useful definition is that "God knows all that *can* be known," which is still consistent with the previous notion whereby God is deemed to be outside time.[17] However, *if* God experiences successive events, as we do, then is our future *already* known to God?[18] Concerning responsible agents, as we shall see, much depends on what we mean by free will.

This topic of God's foreknowledge has become controversial among some Christians. To question the classic notion of God's timelessness, and hence the traditional understanding of omniscience, is perceived as "limiting God" and hence reducing God's glory. But is this really the case? There are certainly biblical passages, especially in the Old Testament, that lend

---

15. Plantinga et al., *Introduction to Christian Theology*, 107. Furthermore: "Time on this proposal is not a metaphysical entity in itself, but simply a measurement of movement or sequentiality." Ibid., 107.

16. The doctrine of creation implies that God transcends time. From a modern physics perspective, time—like space—is a part of creation and therefore dependent upon God. For Alan Padgett, God's timelessness is *relative* to physical time, but he does exist in his own time. William Lane Craig argues that since the creation of the universe God has been temporal, and prior to that, the Trinitarian life is best thought of as being timeless. See Ganssle, *God and Time*.

17. Even this definition is not precise enough—see the exchange in Beilby and Eddy *Divine Foreknowledge*, especially Lane Craig's article and Boyd's reply.

18. Putting it more formally: What then is the ontological status of the "future"? Is it real; is it actualized "now" *for God* in the same way that the present is actualized? Notice that the question is *not* really about the nature of God, (i.e., about what God "knows," for we said earlier "God knows all that *can* be known"), but about the nature of *created* time itself.

support to the view that God knows all the future in detail. But there are also passages that indicate a genuine dialogue between God and his people—a "personal relationship," as Protestants like to emphasize. In some cases God even appears to change his mind in response to those interactions with individuals. This strongly suggests a real relationship, such that history's script cannot be completely written and there is some genuine room for maneuver. I am not going to consider the biblical evidence here, not because I don't think it is important—it is *very* important—but because it has already been addressed extensively elsewhere.[19] Moreover, *sola scriptura* cannot resolve the issue because much depends on one's understanding of biblical inspiration, one's principles of interpretation, and all the contextual factors that shape the way we read and understand scripture—as discussed in chapter 2. None of us approach Scripture with minds that are a blank slate.

What has been the church's *tradition* on the relationship between God's foreknowledge and human free will?[20] There have been two main trends; at the risk of gross over-simplification, one group following Augustine, Thomas Aquinas, Martin Luther, and John Calvin emphasizing predestination (divine determinism), and the other group following John of Damascus, John Chrysostom, Luis de Molina, Jacobus Arminius, and John Wesley emphasizing free will. Moreover, the Protestant debates between Armenians and Calvinists on these matters are mirrored by disputes in the Roman Catholic Church between Molinists and Thomists. As in the case of scriptural interpretation, there is no simple consensus on this subject from the viewpoint of church tradition, and so it is prudent to consider the alternatives graciously.

What, then, can we learn by the use of *reason*? There are various philosophical views on divine foreknowledge; I will briefly mention four of them—and each has its variants.[21] The first viewpoint, "simple foreknowledge," regards God as knowing the future *passively* (as opposed to actively) by virtue of God being outside of time and hence "seeing" all times—past,

---

19. See Sanders, *God Who Risks*; Pinnock, *Most Moved Mover*; Boyd, *God of the Possible*; Pinnock et al., *Openness of God*.

20. This is reviewed briefly in Sanders, *God Who Risks*, 140–72, and Jowers, *Divine Providence*, 11–22.

21. This complex topic has been eloquently and accessibly summarized in Evans and Manis, *Philosophy of Religion*, 42–52, and Peterson et al., *Reason and Religious Belief*, 176–87.

present, and future—at once, as mentioned earlier.[22] Another key assumption is that God's foreknowledge can only arise *after* he has made the decision to create humankind.[23] In other words, God's knowledge of our future is contingent on our existence. At the moment our world was actualized, God timelessly knew *all* of history. Furthermore, it is important to stress that God does not actively *cause* us to choose what we do choose, because he made us with libertarian free will. The choices we make freely imply that we are really able to choose otherwise; we have genuine options. God simply knew (or "knows" since God is in the timeless present)—passively and exhaustively—the outcome of our decisions.

Simple foreknowledge, like all viewpoints, has its critics. Some question the logical coherence of God's timeless knowledge of all events and our genuine free will. It is an important issue, but that aside, one obvious point is simply: What *use* is God's simple foreknowledge *to God* in his providential care for the world? This passive "seeing" of all history does not enable God to do anything different.[24]

So far we have assumed God's timeless knowledge of cosmic history is *passive*—God cannot help but know it. But God can *actively* know all of history if he *planned* it that way. This leads us to the second viewpoint. In classical theism, following Augustine and Calvin, God foreknows all that will occur because he preordained everything. Furthermore, God's omnipotence means that he cannot be thwarted in bringing about his plan—which makes for a strong link between these two divine attributes. Consequently, there can be no surprises for such an all-determining God. Classical theism also affirms God's immutability; we do not affect God and there is no give-and-take in God's interaction with us. Yet this view also holds that human agents are free to do as they desire and are responsible for (at least some of) choices they make. Nevertheless, those choices are all within God's sovereign plan. As God determines or causes all things (either directly or indirectly), he knows all things—from the very beginning. Even if we pray, for example, then God not only scripted the prayer, but has already factored his response into history. In what sense, then, has humankind got free will? Obviously not in the *libertarian* sense mentioned earlier,

---

22. This *timeless* view is linked to Boethius (ca. 480–528). Technically, if God is outside time he does not *fore*know anything! Alternatively, if God is *everlasting*, the essential issues are the same as God only knows passively our decisions after we make them.

23. This is regarded as part of a *logical* ordering in the mind of God, not a *temporal* sequence, since God is timeless.

24. See ibid., 182–85.

On the Nature of God

since we can never choose other than we do. This "free will" is referred to as *compatibilist* in that it is claimed that moral responsibility is compatible with divine determinism.

Can we establish independently what kind of free will we possess—compatibilist or libertarian? Obviously not; it is a faith position. Even so, I—like many—struggle to see how human freedom can be consistent with theological determinism.[25] Moreover, in the context of the problem of suffering, does not divine determinism also make God the author of evil—and hence a "moral monster"? Classical theism's compatibilist claim for moral evil is "no"! Even if that is correct, which I question, what about "natural evil"? God's causation of *all* things, directly or indirectly, as classical theism asserts, has serious theological consequences which cannot be easily brushed aside.

A further view is that of "middle knowledge," or Molinism, which was first formalized by the Jesuit Luis de Molina in the sixteenth century. This ingenious hybrid seeks to maintain genuine human free will and yet hold that all of history is already known to God. How does that arise? God knows not only what *will* come to pass, he also knows what *would* have come to pass if he had chosen to create any other world. Prior to creating this world, God foreknew a wide range of *possible* worlds that could be created. In addition, God is deemed to know what an individual *would* freely choose to do *if* placed in a given situation. These "if-would" combinations are referred to as "counterfactuals" (i.e., *if* placed in this specific set of circumstances, St. Peter *would* deny Christ three times).[26] Now *if* God can place such a hypothetical person in an *actualized* world, and in that exact situation, then God knows what that person *would* freely decide. God is then able to filter the enormously large set of *possible* worlds with the subset of *feasible* worlds in which *all* of humankind's counterfactuals are true. William Lane Craig, a contemporary advocate of Molinism, states:

> Thus, by employing his counterfactual knowledge, God can plan a world down to the last detail and yet do so without annihilating

---

25. This is made more troublesome with the strict Calvinist's doctrine of double predestination. This is the view that God has already determined the eternal destiny of every person. He has chosen some to eternal life and foreordained others to everlasting punishment.

26. Unlike simple foreknowledge, which only "sees" past, present and future as *actualities*, this view claims that God knows counterfactuals of creaturely freedom as true. Simple foreknowledge rejects God's knowledge of counterfactuals as meaningless, since God cannot know as true something that does not exist.

creaturely freedom, since what people would freely do under various circumstances is already factored into the equation by God.[27]

All this knowledge is available to God *prior* to his choosing which one of the feasible worlds to actually make. God then chose to create *this world* because he foresaw that this exact world *best* met his objectives for creation while preserving the freedom of creatures. Indeed, based on God's knowledge of counterfactuals, God selects *whom* to actually create so as to best fit into his overall plan! In making this particular world, God is taking no risks, as he has already examined and rejected all the other feasible worlds. Consequently, there are now no open-ended possibilities.[28]

By God's use of middle knowledge, Molinists assert that God has sovereign control of the actualized world while maintaining humankind's genuine free will. Some question whether counterfactuals even exist for God to know. Others are concerned that their existence limits God's sovereignty over what kind of world he can create. Others ask if humankind's genuine free will is compromised by this pre-selection process and so undermines the personal character of one's relationship to God. Indeed, since the other feasible worlds are—we assume—never actualized, it could be said that the only "entity" with genuine free will is a conceptualized construct within the mind of God and not the real physical person. More troubling, is that—like in classical theism—this world's evil has been deliberately planned by God. As Thomas Flint puts it:

> If Judas sins, it is because God knowingly put him in a set of circumstances in which he would sin, and knowingly refrained from putting him in a set of circumstances in which he would act virtuously.[29]

Deeply disturbing though that thought is, perhaps a redeeming feature is that this world is the *best possible* for Molinists.

The fourth option, relational (or open) theism, emphasizes genuine human free will and a God who, in a meaningful sense, functions temporally—at least since the creation project began. God *relates* with his creation, especially humankind, and is affected by that relationship. From this perspective, God has exhaustive knowledge of the past and the present,

---

27. Lane Craig, "Middle Knowledge," 122.

28. Beilby and Eddy, *Divine Foreknowledge*, 11, emphasis mine. See also Peterson et al., *Reason and Religious Belief*, 178–81.

29. Cited in Peterson et al., *Reason and Religious Belief*, 181.

and certain elements of the future—the things that do not involve free will, such as the confidently predictable physical processes in the cosmos. The future, though, is not completely fixed (i.e., "closed" or pure mechanism) but "open" to what both God and humans decide to do. And, as we saw in the last chapter, there is also an element of "openness" inherent within the natural processes of God's creation. There are, therefore, numerous possible futures and not just one fixed Master plan—God is flexible. I find Jürgen Moltmann helpful in this regard:

> When God restricts himself so as to make room for his creation, this is an expression not of powerlessness but of almighty power. Only God can limit God.... But where he has created something, God respects the unique character and the liberty of what he has created.... If this self-limitation is true of his omnipresence, then it is also true of his omnipotence and his omniscience. God has created beings with relative independence. By limiting his omnipotence, he has conferred the free spaces their freedoms require. Because of the restriction of his omniscience he cannot foresee how those he has created will decide, and how they will develop. He leaves them time, and opens up for them an unforeseeable future.... He learns from them.[30]

Consequently, God knows as possibilities and probabilities those events which *might* happen in the future, one of which will be *actualized* once an individual freely makes their choice.[31] In the cosmic project that God has in mind, God anticipates what we will do—and he can have all sorts of plans based upon what we *might* do in any situation and then he is ready to respond according to what we actually do. God, together with responsible creatures, continually creates the future—or, rather, the present—as history goes along. God's "knowledge" is therefore dynamic; it changes with time as potentialities become actualities. Given the ever growing population and

---

30. Moltmann, *Science and Wisdom*, 120. Clark Pinnock also writes: "God uses omnipotence to free and not enslave.... In a sense, God limits his own power to allow us to be free. The power itself is unlimited but God chooses to actualize a particular world whose development he leaves largely in the hands of creatures. This leaves the future open and largely under our control. God does not predetermine it, although he could do so. Does this idea diminish omnipotence or enhance it? What other than omnipotence could create free creatures and still feel confident that its purposes will be realized?" See Pinnock, *Most Moved Mover*, 94.

31. In the language of Molinism's counterfactuals, God knows "if-might"—rather than "if-would"—counterfactuals (e.g., *if* Peter was placed in a given scenario, he *might* deny Christ three times.)

the myriad of choices that individuals make—many of which affect others—this scenario is a matrix of interconnected decisions of ever increasing complexity. But a God of *infinite* intelligence can handle this matter! Note too that *if* it were possible to rerun time again, not only would evolutionary history be different, but we would be able to decide differently—a possibility that cannot arise if the future is already fixed in the mind of God. For God to know the outcome of our decisions before we actually make them would destroy the very freedom he gave us. It would destroy the possibility of genuine *love* and true *faith*.[32]

Relational theism, not surprisingly, is criticized by those who regard God's absolute control as an essential feature of God's sovereignty. In other words, if God is not in *total* control, he cannot ensure the outcome of history (i.e., that evil will ultimately be defeated).[33] Following from that, are the risks God is taking morally defensible? The obvious response is that God, being morally perfect, is not a reckless gambler but is willing to create for the sake of *love*—and love, of necessity, entails the risk of rejection. A further question is whether God's transcendence has been abandoned completely in favor of his imminence, and if not, where and how is transcendence manifest?[34]

Controversy and debate abound on such matters—and there is much nuance that I have glossed over.[35] However, we must also remember that philosophers are interested in logical consistency and conceptual possibilities, not necessarily what is really—or even probably—the case. The "god-of-the-philosophers" has always been different from the revealed God of the Judeo-Christian tradition. Yet philosophical theologians are mindful of religious sensibilities—such as providence. And coherent insights arising

---

32. Polkinghorne also favors a similar temporal, relational view; see Polkinghorne, *Exploring Reality*, 118–19. He also points out that atemporal and temporal views of God can be informed by science, but in the final analysis this issue can only be settled by a philosophical decision (ibid., 114).

33. Relational theism is a tree of many branches, one of which (process theology) does not guarantee God's victory over evil. Other branches, such as open theism, respond by saying God's power is more than sufficient to fulfill his purposes.

34. Another understandable concern is the matter of prophecy in Scripture. For a full response, see footnote 19 of this chapter.

35 For further details, see, e.g., Jowers, *Divine Providence*; Ware, *Doctrine of God*; Basinger and Basinger, *Predestination and Free Will*; Hall and Sanders, *Does God Have a Future?*; Cobb and Pinnock, *Searching for an Adequate God*; Tiessen, *Providence and Prayer*; and Hasker, *God, Time and Knowledge*.

from such thoughtful deliberations are valuable to Christians and enhance our faith in God and his actions in history.

In summary, God is inconceivably creative, infinitely intelligent, and has unsurpassable foreknowledge. He is omnicompetent and can achieve his goals. All four perspectives on God and time agree with those assertions. The debate is in the details, on how, when, and what God knows (and can know), and on whether God is timeless or everlasting. All the views have perceived strengths and weaknesses, from the eyes of both reason and faith. All wrestle with God's transcendence and immanence. As Evans and Manis conclude:

> There is no easy solution to the problem of foreknowledge and freedom, and whichever solution one adopts will have significant repercussions for one's broader theology. . . . There is a price to be paid for each solution.[36]

SUMMARY AND CONCLUSION

In classical theism there is a tight weave in the fabric of divine attributes. As we have seen, that fabric begins to unravel when you pull on the threads of immutability for *theological* reasons. Furthermore, while Christian theology is informed by reason, it should not be held hostage by philosophy, particularly if that is heavily influenced by Hellenistic thought.[37] It is, of course, inevitable that theology is influenced by the surrounding culture. The early and rapid spread of the gospel in the Greco-Roman world was, in part, because it *could* be readily adapted to the prevailing worldview. While I think that synthesis was providential, nevertheless, it has a legacy of which we must be critically aware. This is nothing new; Pinnock reminds us:

> We are somewhat in Blaise Pascal's position when he realized how different the God of the philosophers was from the God of Abraham, Isaac, and Jacob. Pascal recognized that there was a pagan legacy, which needed to be overcome. The tendency was to present God as one-sidedly transcendent—completely separate from and above the world—and not as the living God. It is a tendency embedded in our own thinking and leads us to suppose that it is biblical when it is not. . . . Jesus spoke Aramaic, not Greek, and

---

36. Evans and Manis, *Philosophy of Religion*, 51.

37. Plantinga et al., *Introduction to Christian Theology*, 84–102; Pinnock, *Most Moved Mover*, 65–111.

the Bible was written in Jerusalem not Athens. The doctrine of God was, however, shaped in an atmosphere influenced by Greek thought.... Tertullian asked a famous question: "What has Athens to do with Jerusalem?" His query implied it had nothing to do with it, but the answer as concerns the doctrine of God is actually "much and in every way"![38]

There will, in the final analysis, always be an element of mystery. That is why humility is required in defending our passionately held views and grace extended to those who see things differently. But we should not invoke "mystery" prematurely, or use it to wrap up confused, incoherent ideas and demand others accept them "by faith." Concerning "mystery," where do we locate it? In the nature of *God*? In the complex character of *creation*? Or in God's multifaceted *relationship* with creation? I think it is helpful to at least distinguish between these sites of mystery. Despite the tremendous strides scientists have made in understanding nature, since we are finite and fallible, there will always be an element of mystery about the cosmos *for us*. But that is not the only locus of mystery. Migliore's quote at the beginning of the chapter rightly reminds us of the inevitable mystery of God. What we know about God is founded on faith—from trusted beliefs as part of a faith community and based on our individual encounters with God. Our understanding of God is always *interpreted* truth, based on God's relationship with his creation as Creator, Reconciler, and Sanctifier. Theology's relationship with science is understood solely in terms of God as Creator (and Sustainer), so allowing for an understanding of *general* providence. Jesus as Reconciler and the life-giving Spirit as Sanctifier are both personal, and leave space for *particular* providence—which will be considered in the next chapter. But having identified three sites of mystery, each one remains a riddle. Nevertheless, as Christians we live in both faith and hope. Recalling the words of Paul: "For now we see in a mirror, dimly, but then we will see face to face. Now I know only in part; then I will know fully, even as I have been fully known" (1 Cor 13:12).

---

38. Pinnock, *Most Moved Mover*, 66, 68–69. He further quotes Donald Bloesch: "A compelling case can be made that the history of Christian thought shows the unmistakable imprint of a biblical-classical synthesis in which the ontological categories of Greco-Roman philosophy have been united with the personal-dramatic categories of biblical faith." Ibid., 68–69.

Chapter 7

On Miracles and Prayer

> It may be the case that the natural world is more complex and more open than we ordinarily assume, and that the network of causal connections is less restrictive than we imagine.
> —Peter Baelz[1]

INTRODUCTION

This study of the relationship between science and Christianity has so far explored the role of God as Creator and Sustainer of the universe. This "big picture" view can be referred to as God's *general* providence. This sentiment is expressed by Jesus in the Sermon on the Mount: "[God] makes the sun to rise on the evil and the good, and sends rain on the righteous and on the unrighteous" (Matt 5:45). But Christianity maintains that God is *relational* and therefore intimately involved in the lives of both individuals and communities. In the Old Testament, the Jews understood themselves as worshiping the relational God of Abraham, Isaac, and Jacob. In the New Testament, Jesus taught his disciples to pray "Our Father . . ." (Matt 6:9–13), again emphasizing relational intimacy with the divine. And the gospel accounts call attention to the personal encounters that Jesus had with individuals throughout his ministry. While general providence may lead some to think—with the deists—that God is "hands off," the biblical story

---

1. Baelz, *Does God Answer?*, 52.

emphasizes a God who is "hands on." Any serious dialogue between science and Christianity must therefore recognize God's covenantal commitment with humankind (indeed, the whole of creation). It is this personal commitment, most graphically demonstrated in the incarnation, which continues to inspire the devotion of Christ's followers. Two key features in Christian experience are the life of prayer and occasional events that are believed to be miraculous. In this chapter we will briefly explore these topics from both scientific and theological perspectives.

MIRACLES

The Bible, like most religious texts, contains many accounts of miracles. Yet, are miracles plausible in a scientific age? If it is unreasonable to believe that such "supernatural" events can occasionally occur, then—following Bultmann—we will need to demythologize them.[2] Before we address this matter, it is worth stating that one can accept the possibility—even reality—of miracles without being coerced to believe that every miraculous event in the Bible is historically accurate. For example, I don't consider the miracle where "the sun stood still in the sky" during Joshua's epic battle to have occurred at all (see Josh 10:12–14). Omitting a discussion on the immense astronomical implications of such an event, if it had occurred, it would have had global consequences and so probably would have been recorded in every major culture's literature. The Egyptian and the Mesopotamian civilizations, for instance, were very keen on astronomy and would have noted such a dramatic event. The incident does, however, depict Joshua as having had a victory of legendary proportions and worthy of suitable folklore. This goes to show that *some* of the biblical accounts of the miraculous are better understood as *literary* rather *literal* history.[3] In the New Testament, there is another peculiar story of the coin that Jesus predicted would be found in the fish's mouth in order to pay Peter's and his own temple tax

---

2. Evans and Manis, *Philosophy of Religion*, 118.

3. Biblical scholars have long noticed the difference between the books of Judges and Joshua in how they depict the relationship between the young nation of Israel and their Canaanite neighbors. In contrast to Judges, Joshua's epic battles give the impression that the inhabitants have been completely driven from the land. Furthermore, there is no archeological evidence for Joshua's famous battle over Jericho—and many of the other reported victories. All this gives important insight and context to the historicity of the book of Joshua and the role of meaning-making storytelling in the Old Testament. See, e.g., Enns, *The Bible Tells Me So*.

(Matt 17:24–27). Commenting on this gospel passage, which is unique to Matthew and seems more fable than fact, Eugene Boring concludes:

> Taken literally, the story has problems not only of physics but also of ethics, and it conflicts with other pictures of Jesus, who does not use his miraculous power for his own benefit.[4]

In other words, in addition to the scientific scrutiny of any given miracle, we need to discern the *theological significance* of the event, i.e., is this miracle consistent with the overarching biblical storyline? We will return to this matter later.

While we must use critical reasoning in exploring Scripture, Christianity cannot escape the existence of the miraculous at the heart of its story. In particular, that in the resurrection God raised Jesus bodily from the dead. Everybody knows in all cultures throughout history that dead people do not come back to life. Nevertheless, I have come to believe that this event did happen. It is Christianity's foundational data point and one on which the whole of cosmic history turns. This is not the place to explore the evidence for my belief in the resurrection, which is shared with countless others in orthodox Christianity. Rather, let us focus on the dialogue between science and Christianity on this general topic, beginning with clarification on what we mean by a "miracle."

Some Ways of Defining "Miraculous"

As we saw in chapter 5 with the term "chance," a "miracle" also has a multitude of popular connotations. One is a *totally unexpected event*, such as a horrific car crash which somehow leaves a baby completely unharmed, or a coincidental meeting with a long-lost friend while on holiday abroad. In neither of these examples was prayer involved, and so the events cannot be attributed as God's *response* to prayer. Consequently, it is truly a surprising or unanticipated event. Even so, those with a religious inclination may, with hindsight, attribute the unlooked-for event as a special act of God. Nevertheless, in both cases, no one is necessarily claiming that God has "intervened" with the laws of nature to bring about the event.

Miracles, however, are usually considered in a religious context and interpreted as being a result of some kind of *request* for divine activity. Although the birth of a healthy baby is perfectly natural, in a biological sense,

---

4. Boring, "Matthew," 372.

the wonder of new life is often acclaimed as a "miracle." Consider too the "miraculous escape" of a family emerging unscathed from a storm cellar to discover that their house has been demolished by a tornado. In contrast to the previous examples, let us say in both these cases prayers were made (i.e., for a safe delivery of a healthy baby and to survive the storm). In both situations God's general provision, via natural processes, has been continuous throughout and so on one level there has been no apparent change in divine activity. Yet these *particular* people prayed genuine prayers with some expectation that God had the capability to (somehow) bring about their desired outcomes. They attribute their *specific* outcomes to divine activity, no doubt enhancing their faith in God's special providence. Nevertheless, some theists might not want to describe those outcomes as "miraculous," because they are compatible with natural laws. Instead, they might prefer to articulate the outcomes in terms of God's ever-present sustaining activity.

While such theists are logically correct (depending on your definition of "miracle"), there is a further issue when going from the *general* to the *particular* case. We do not know the *context*. While God is emphatically not the author of evil, we are aware that the desired, positive outcomes of both those scenarios are not guaranteed. Let's say that the parents had been desirous of a child for some time and they already experienced two late miscarriages and a stillbirth. Does that change the situation? Those who live in tornado alley are aware of this annual season and the devastation it causes. Perhaps the family had experienced the loss of other family members or close friends to previous tornadoes. In both cases their personal history informs their prayers. While the theist can still maintain that God has not necessarily "supernaturally intervened" in these two cases, one does not want to fall into the reductionist's trap and simply think we have therefore explained the events "away." One would be hard pressed to claim that these outcomes are *not* a sign of God's presence and special providence. Certainly the people involved have grounds to thank God and, if they do so, they would interpret the event as God revealing himself to them in a uniquely special way. Such events are interpreted as acts of God *through the eyes of faith*. They are examples of *subjective* miracles, even if they are not rigorously *objective* miracles. This calls for a broader definition of "miracle."[5]

More typically in popular culture, however, miracles are viewed as a *direct* act of God and often, but not exclusively, in connection with prayer. A working definition of a miracle is an event that would not have occurred

---

5. See Evans and Manis, *Philosophy of Religion*, 128.

in the exact manner that it did if God had not *directly manipulated* the natural order.[6] This is consistent with the view of Aquinas. He regarded a miracle as a surprising or puzzling work of wonder; God bringing about something which nature's innate powers could never do.[7] Miracles occur by God temporarily elevating nature's powers beyond that which is normal. Consequently, miracles have their ultimate origin in God alone.[8]

Miracle as "Violating" the Laws of Nature

As indicated above, a miracle today is often viewed as an *intrusion* into the otherwise smooth running of the cosmos. This view is typically associated with the eighteenth-century philosopher David Hume, who asserted:

> A miracle is a *violation* of the laws of nature; and because firm and unalterable experience has established these laws, the case *against* a miracle is—just because it is a miracle—as complete as any argument from experience can possibly be imagined to be.[9]

Polkinghorne muses that "it is rather touching to find the stern critic of induction placing such firm faith on the unalterable character of nature's laws."[10] Keith Ward reminds us that Hume uses language that is consistent with a clockwork universe, a closed physical system working in a wholly deterministic way.[11] Consequently God can only act in such a system by breaking some of its own laws and *interfering* with it. Since—presumably—a perfect Creator is capable of designing a non-defective clockwork cosmos, it would be absurd to think that God would need to intervene to correct a malfunction. Consequently, any interference would be deemed as either irrational for—or arbitrary meddling from—the Creator.[12]

---

6. Adapted from Peterson et al., *Reason and Religious Belief*, 194, emphasis mine.

7. Aquinas, *Summa Contra Gentiles*, 101, 474. I am here referring to Aquinas's miracle of highest rank.

8. Ibid., 102.2, 476.

9. Hume, *Enquiry concerning Human Understanding*, sec. 10, pt. 1, 58, emphasis mine.

10. Polkinghorne, *Science and Providence*, 63.

11. Ward, *Divine Action*, 179. Elsewhere Hume writes: "God, *who first set this immense machine in motion* and placed everything in it in a particular position, so that every subsequent event had to occur as it did, through an *inevitable necessity*...." Hume, *Enquiry concerning Human Understanding*, sec. 8, pt. 2, 50, emphasis mine.

12. Ibid., 179.

With the demise of the mechanistic view of the cosmos, nature's "laws" are not as rigid as Hume thought—as we saw in chapter 5. In addition to the repetitive patterns of nature, the universe contains—as Ward puts it—"unique events and processes, emergent states and surprising sequences of probabilistic causal interconnections."[13] Moreover, with the logical problem of induction, the laws of nature can never be verified or proven true, which contributed to Popper's introduction of his falsification principle—as discussed in chapter 3.[14] Since laws are *descriptive*, rather than prescriptive, it is misleading to describe God's action in deviating from them, on occasion, as a "violation of a law."[15] As Polkinghorne points out:

> The problem of miracle is not strictly a scientific problem, since science speaks only about what is usually the case and it possesses no *a priori* power to rule out the possibility of unprecedented events in unprecedented circumstances.[16]

Nevertheless, it is still fair to say that a miracle is an exception to the normal processes of nature.

Regardless of the status and scope of scientific "law," Keith Ward makes an astute observation that there is still a logical space for miraculous events in the way that Hume defines it. This is because any general structure can stand some exceptions without being destroyed as a structure. But those exceptions must be very rare and, therefore, seen as *anomalies*.[17] Evans and Manis also point out that if an event occurs which appears to contravene a law of nature, two options are possible. We can regard it as evidence to falsify that existing summary statement that describes the normal way nature behaves, or we can regard it as an exception to the rule that otherwise

---

13. Ibid., 179.

14. In addition, it may be that some particular, rare event is presently deemed to be a miracle but it may not be truly "miraculous" because there may be physical principles to explain the occurrence that we have yet to discover. This possibility will always exist and can never be overlooked. Holding out in hope, however, that such future "natural" explanations will, eventually, eliminate the notion of the miraculous is futile, due to the humankind's inherent finiteness. Nevertheless, I suggest that all this concern is merely a distraction and arises because it is overly focused on Hume's narrow definition of "miracle" and his assumed character of both physical law and the cosmos.

15. See Evans and Manis, *Philosophy of Religion*, 126. They discuss Hume's arguments at length—see 125–35. For a more detailed study of Hume on miracles, see Brown, *Miracles and the Critical Mind*, 79–100.

16. Polkinghorne, *Quantum Physics and Theology*, 35.

17. Ward, *Divine Action*, 170–1.

holds. The former is reasonable when the evidence for that exception is *repeatable*, meaning that if the precise circumstances were duplicated, the exception would occur again. In such circumstances, the law would need to be revised; indeed that is how science progresses. And we would not describe the event as miraculous. But if it were—as far as we can ascertain at this point in time—an *unrepeatable* occurrence, it would seem irrational to abandon the belief in the law of nature that holds in every other case. Evans and Manis conclude:

> There is no compelling reason to use the phrase "laws of nature" only to describe laws that hold without exception. . . . It seems rash therefore for philosophers or others to claim dogmatically that miracles *cannot* happen. Miracles seem *possible* at least, and it also seems possible for there to be compelling evidence for their occurrence—evidence of the ordinary historical kind.[18]

In summary, the laws of nature are based on the regular patterns we observe, whereas a miracle by definition is unprecedented, unexpected, and unrepeatable. Despite this brief rationale from a scientific and philosophical perspective, the influence of determinism lingers on and many are still uncomfortable with invoking "supernatural" explanations. Mackay is helpful on this point: "For Biblical theism, the miraculous is not so much an intervention (since God's sustaining activity is never absent) as a *change of mode* of the divine agency."[19]

Exploring the Theological Significance of the Miraculous

If the laws of nature are the "habits of God" then what is the *theological* rationale for God to change his customary mode of behavior to do something that is radically out of the ordinary? Polkinghorne, Ward, and others seek coherence between a specific miracle's *theological* significance and maintaining an inherent rationality to God's modes of activity.[20] This is a

---

18. Evans and Manis, *Philosophy of Religion*, 135, emphasis mine. Peterson et al. make exactly the same point; Peterson et al., *Reason and Religious Belief*, 195.

19. MacKay, *Science, Chance, and Providence*, 18, his emphasis.

20. See Polkinghorne, *One World*, 76. Ward writes: "It is quite unsatisfactory to think of miracles as just rare, highly improbable and physically inexplicable events. The theist has no interest in the claim that anomalous physical events occur. Events in which the theist is interested are acts of God; and Divine acts do not occur arbitrarily, or just as anomalous and wholly inexplicable changes in the world. They have a rationale; and

key feature for Polkinghorne, to the extent that he claims there is no sharp separation between God's general and special providence:

> The discontinuities which the language of natural and miraculous suggests, or the divisions between God's types of will, are matters of human convenience, relating to the differences in our perception and not to fundamentally distinct kinds of activity in God.[21]

Polkinghorne makes an analogy of the familiar observation, whereby the steady increase of temperature suddenly produces a *discontinuous* change from liquid to gas at water's boiling point. No one claims the underlying laws of nature have changed at 100°C at this dramatic phase transition.[22] Evidently, a change of circumstances can result in very different modes of behavior. Polkinghorne is at pains to avoid the language of "intervention," "exception," and even "unrepeatable," since they smack of arbitrariness in God's rationality.[23] He points out that if *natural* phenomena manifest such radical unexpectedness, need we distinguish between God's modes of activities? It is a fair point in the context of intervention, although there remains the issue of repeatability. Polkinghorne concludes that the fundamental theological problem of miracle is

> how these strange events can be set within a consistent overall pattern of God's reliable activity; how we can accept them without subscribing to a capricious interventionist God, who is a concept of paganism rather than of Christianity. Miracles must be perceptions of a deeper rationality than that which we encounter in every day, occasions which make visible a more profound level of divine activity. They are transparent moments in which the kingdom [of God] is found to be manifestly present (Matt 11:2–6).[24]

While Polkinghorne is right to emphasize God's steadfast faithfulness, there remains a pertinent theological question: "Can God change his mind?" Responses to this question are inevitably wrapped up in language

---

that rationale must be connected with the purposes of God for the world." Ward, *Divine Action*, 176.

21. Polkinghorne, *Science and Providence*, 59.

22. Ibid., 60.

23. Ibid., 61. He avoids the contentious arbitrary element of "unrepeatability" by pointing to the subtle complexity in human events such that circumstances can never actually be repeated. It should be noted that process theology also rejects language of divine "intervention" and other terms that imply supernaturalism.

24. Ibid., 60.

On Miracles and Prayer

of God's engagement with *time*, which was considered in the previous chapter. I suggest this adds a significant complexity to the general question of miracle and divine activity. This is a timely—and contentious—theological issue, one that is equally relevant to the matter of prayer.

I have, however, a further niggling suspicion that Polkinghorne is too cautious. A key feature that special providence emphasizes, over general providence, is the *personal* nature of God. God's acts are, to use anthropomorphic language, a coherent combination of head *and* heart. Our focus on divine rationality downplays the latter. From a Trinitarian perspective, Jesus's ministry reveals the heart of God the Father. In light of that, there is enough space between "character" and "action" for a God of love to do unprecedented acts without compromising divine integrity or the overall eschatological goal of history. In other words, there has to be room for the possibility of grace—sheer grace—which defies rationality. Perhaps God's grace is in Polkinghorne's category of "deeper rationality"!

Moving on, however one construes the historicity of the gospel accounts, we recognize Jesus did not heal everyone in need within the geographical regions he covered during his earthly ministry, but neither did he heal nobody. I am sure the woman who was suffering from hemorrhages for twelve years (Mark 5:24–34; Luke 8:43–48; Matt 9:20–22) was not the only person in need of a miracle in that crowd. Performing miracles was a feature of his fame (and hence should not be treated as mythical). N. T. Wright is insistent:

> We must be clear that Jesus's contemporaries, both those who became his followers and those who were determined not to become his followers, certainly regarded him as possessed of remarkable powers. The church did not invent the charge that Jesus was in league with Beelzebul (Matt 12:24–32; Mark 3:20–30; Luke 11:14–23); but charges like that are not advanced unless they are needed as an explanation for some quite remarkable phenomena.[25]

In that vein, Richard Swinburne speaks of miracle as an event of an *extraordinary* kind of *religious significance*.[26] That being the case there is a *subjective* element to the miracle, not just the *objective* nature of the event itself.[27] Ward, who—unlike Hume—sees the universe as an open system, writes:

---

25. Wright, *Jesus and the Victory of God*, 187. For further appreciation of exorcisms, see 195–96.

26. Swinburne, *Concept of Miracle*, 1.

27. Ward, *Divine Action*, 177.

> Miracles are not just anomalous events which interrupt the seamless processes of nature. . . . They are not merely physically inexplicable events, but astonishing and spiritually transforming signs of Divine presence, purpose and power. God brings miracles about by a special intention to enable creatures to come to a more conscious and dynamic relation with him. . . . A miracle, as an extraordinary act of God, essentially has the character of a communication, possessing an intended meaning which is to be discerned by those who apprehend it in faith.[28]

This broadens the definition of "miracle." Too often the focus is on the physical event and not the *meaning* of the event.

In the Bible, a miracle functions as a *sign*, a communication of God's relationship with humankind and history.[29] The most common reference to a miracle in the New Testament is the word *dynamis*, meaning "a mighty work" or a "deed of power." Another Greek word simply corresponds to that of "sign." In both cases these words do not have the connotation of a "supernatural" act that is typically assumed today.[30] Instead, they point to God being at work in unexpected, powerful, and significant ways, which seem—as N. T. Wright puts it—"to provide evidence for the active presence of an authority, a power, at work, not invading the created order as an alien force, but rather enabling it to be more truly itself."[31] If the mighty deeds of Jesus are signs, what do they signify? They point to God powerfully at work, inaugurating the long-awaited time of liberation that is a feature of the kingdom—or reign—of God (see Luke 4:18). They are therefore *signs of the presence of the reign of God*, or, using a Jewish term, of *shalom*.[32] This is very different from pointing to Jesus as an egocentric magician conjuring tricks to instill faith or prove his identity.[33] Neither was he using his powers for his own benefit or whimsically dispelling random acts of kindness.

---

28. Ibid., 180.

29. Cotter, "Miracle," 99. The word "miracle" comes from the Latin word *miraculum* meaning "marvel." Yet this word is never used in the fourth-century Latin Vulgate.

30. Wright, *Jesus and the Victory of God*, 188.

31. Ibid., 188. We must therefore be mindful that our reading of the biblical texts is not colored by the Enlightenment language and connotations of "supernatural."

32. See ibid., 190, 192–93.

33. Neither should they be seen as primary evidence for the *divinity* of Jesus. While the seeds of the divinity of Christ are present in the gospel accounts, especially the 4th Gospel, the formulation of that doctrine emerged later in the church councils. The Synoptic Gospels, in particular, emphasize Jesus as the long-awaited *Messiah*.

Moreover, *shalom* implies wholeness of relationships between humankind and God, with each other, and with *creation*. Within such Jewish expectations, mighty deeds that involve the natural world are only to be anticipated. Such deeds were evidently recognized by the gospel writers as signs—since healings and "nature miracles" were seamlessly recorded in their accounts.

C. S. Lewis is right when he says that the power behind both creation and Christ's miracles is *personal*:

> The miracles [of Jesus] . . . are a retelling in small letters of the very same story which is written across the whole world in letters too large for some of us to see. . . . In other words, some of the miracles do locally what God has already done universally: (i.e., the miracle of creation) others do locally what He has not yet done but will do (e.g., resurrection of the dead, creation restored). In that sense, and from our human point of view, some are reminders and others are prophecies.[34]

Those "prophetic" miracles of Jesus were also signs, but this time pointing toward the future hope of Christians, when the reign of God comes in all its fullness. In that sense, they serve as a foretaste signaling there is much more to come. It may be unwise to try and categorize each of Jesus's miracles in this way, and hence this distinction is perhaps not too helpful. Nevertheless, Lewis reminds us that the miracles were both personal to the recipients (and their families and communities) *and* part of a bigger narrative—from the creation to the eschaton.

Since the signs point to the presence of the paradoxical "now and not-yet" kingdom of God, rather than vainly toward Christ himself, there is no theological reason to think that God has changed his mode of activity. God is still at work today; we continue to be in the "last days"—as New Testament writers put it. It seems, therefore, only plausible to at least *expect* miracles at the "cutting edge" of kingdom of God activities—where Christ's good news confronts the powers of evil.[35] Christ's earthly ministry confronted the systemic evils of the day, following in the footsteps of earlier prophets. Moreover, Jesus highlights his miracles as evidence of God at work to John the Baptist's disheartened followers (Luke 7:20–23; Matt

---

34. Lewis, *God in the Dock*, 29, see also 32–33; text in parentheses is stated in context from his essay. For further information of Lewis's views of miracles, his terminology, and context see Brown, *Miracles and the Critical Mind*, 229–38.

35. C. S. Lewis astutely observed: "miracles [tend to occur in] areas *we* naturally have no wish to frequent." Lewis, *Miracles*, 274, emphasis mine.

11:2–6). Nevertheless, miracles will be inherently *rare*—by definition—but clusters at a given time and place may well point to a heightened presence of the reign of God, i.e., God's glory.

God, however, does not want us to be miracle-chasers. Fascination with signs and wonders is nothing new. The Jews that Paul encountered evidently asked for miracles (1 Cor 1:22–24).[36] Matthew states that miracle-chasing was a feature in Jesus's ministry too (e.g., Matt 12:39; 16:4). Curiously, perhaps, two of the famous temptations of Christ are to perform miracles for Jesus's personal benefit and to elicit—even compel—devotion.[37] Since Jesus resisted these temptations, we too should avoid indulging our fascination with the miraculous for inappropriate motives.

Summary and Conclusions

Colin Brown speaks of Christian apologists as broadly being in two camps concerning the matter of miracles. The first is an "offensive" group who see miracles as irrefutable evidence of divine intervention and hence the existence of miracles is seen as objective grounds for faith. The second camp is more numerous and is classed as "defensive." While this group thinks miracles are logically possible, that does not mean every reported biblical miracle has sufficient "hard" historical evidence to back it up.[38] Furthermore, American Evangelicalism is traditionally associated with the offensive group, whereas the British apologists are more typically in the defensive camp.[39] Clearly my approach is more in keeping with the latter group than the former, since it also regards miraculous claims in the context of an overarching theological view of God's cosmic purposes.

In summary, we must recognize the pervasive remnant of a clockwork universe that can still color our instinctive reaction toward miracles. Commenting on that inbuilt bias, C. S. Lewis points out:

---

36. Note that this letter is commonly thought to have been written *before* the gospel accounts.

37. Matt 4:1–11 and Luke 4:1–13. Matthew and Luke may well have been alluding to—or making parallels with—Israel's wandering in the wilderness for 40 years. The contrast is stark and theologically significant: Jesus was faithful to God in his trials, whereas the people of the exodus were not.

38. Brown, *Miracles and the Critical Mind*, 197.

39. Ibid., 197, 219. There are also nuanced differences between Protestant and Roman Catholic apologists. For a full discussion, see ibid., 197–238.

> Whatever experiences we may have, we shall not regard them as miraculous if we already hold a philosophy which excludes the supernatural.[40]

Instead, the ardent skeptic will seek to explain them away as illusions, or psychosomatic healing, or—in the case of biblical miracles—as literary devices, or pandering to the ignorant, gullible, and uncritical. As we have seen, there is no logical reason to deem miracles impossible. But that does not mean that we should be naively accepting of all miraculous claims, either contemporary or those recorded in history. Where possible, we should use all the tools of science and medicine to scrutinize those claims. We should also recognize that, from a Christian perspective, a miracle is a gracious act of God of religious significance and a vivid sign of the presence of God's reign.

Furthermore, we must also be mindful not to get caught up in the sterile rhetoric surrounding the notion of "supernaturalism." From God's perspective, *all* his acts are *natural* to him! Miracles are not divine *intervention*, for God is continually and intimately involved in his creation. Consequently, such rare, unexpected, unrepeatable events point to the presence of the reign of God in power. The concern for a theological rationale that defends God's integrity is justified, but should not handcuff God from acting in novel ways—else we also deny the incarnation.

Of greater theological concern to me are two issues: (a) why X was the recipient of a miracle, but not Y, and (b) why are they not occurring more frequently? If miracles are evidence of the "now and not-yet" reign of God, why are there not *more* of them? Given the needs of our world I am, at times, deeply puzzled—to say the least—by the lack of miracles in the face of moral evils (deeds done by human agency) and natural evils (natural disasters and disease). This is a serious theological issue, aspects of which will be alluded to in the next chapter, but the problem of evil is another project for another time.[41] Concerning the first question, I recognize within the gospel accounts, at least by implication, that each miracle is tailored to the individual circumstances in a unique way. The historical particularity of a miracle is therefore inherent in Jesus's apparent "favoritism" as he inaugurated the kingdom of God. This troubling fact is not easily erased away

---

40. Lewis, *God in the Dock*, 25.

41. Recommended reading includes: Rice, *Suffering and the Search for Meaning*; Hasker, *Triumph of God Over Evil*; Wright, *Evil and the Justice of God*; Davis, *Encountering Evil*; Hall, *God and Human Suffering*; Long, *What Shall We Say?*

and theologians are continually challenged to wrestle with the matter in the search for coherence with the Trinitarian character. Rather than perceiving such special treatment as somehow "unfair" on the rest of humankind, I suggest we should see it as the logical inevitability of the Christian claim that God is *personal*.[42] On the one hand this is encouraging and hence we should always be prepared to humbly ask God for a miracle. But experientially—since miracles are so rare—this can also be disheartening, as we are naturally inclined to think we too would be a worthy recipient for one! I am sure many of us feel in the position to echo the words of the father in Mark 9:23 who said: "I believe, help my unbelief."

Perhaps the lack of miracles, at least in the Western world, is—in part—because of the legacy of science and its interaction with the church. Maybe too many of us would use miracles in an inappropriate way, such as a means of "proof," or because we are self-absorbed and don't view them in the broader context as signs pointing to the reign of God. Who knows— and does it really matter? After all, we are not miracle-chasers, are we? In conclusion, I again quote Polkinghorne:

> Too glib an evocation of special providence may trivialize God's action in the world, but the rejection of all such particular action reduces God to an impotent spectator. The religious mind strives to maintain some balance.... The paradoxes of providence are not mere intellectual puzzles. They arise from the heart of religious experience.[43]

Speaking about balance and religious experience, knowing in our core being that "God's grace is sufficient for you" (2 Cor 12:9), together with the fact of Emmanuel ("*God with us*"), is surely enough for our journey.

PRAYER

Prayer too is complex! It contains several facets, such as praise and worship, listening and meditation, as well as bringing our needs and desires before God. It is the Christian's dynamic communication with God. And through prayer we ask and receive forgiveness and wholeness from him.

---

42. Polkinghorne writes: "Total impartiality would be total impersonality—which is not to say that a personal God has to have favourites, but that he will treat particular people in particular ways . . . without the special providence, the idea of a personal God is emptied of content." Polkinghorne, *Science and Providence*, 48–49.

43. Ibid., 51.

On Miracles and Prayer

Not surprisingly, it is the *petitionary* prayer aspect that is the topic of this section. What is it that God does when we articulate such prayers?[44]

Naturally, the importance of prayer, in the life of both the faith community and the individual, means that much has been written on this subject. But it is still a mystery and always will be. Questions on the necessity and effectiveness of prayer will never receive a perfectly satisfactory response, but they still need to be discussed. Why? Because we need to have faith—or confidence—in prayer if we are to practice it. The infrequency of our prayers may indicate that we only ask God as a desperate last resort. If, however, we *say* we believe in a personal God and we *don't* pray we are, in effect, saying "I don't *need* you," or that "I don't believe you have either the *power* or the *will* to act." Mackay is right when he concludes: "A Christian can indeed justly argue that to *refuse* to ask God's help in trouble would make it *irrational* for him to expect to receive it."[45] Thankfully, the Christian view of prayer is far more positive than this: prayer has value because God *invites* it. Prayer is not asking God for a favor, but an expression of our relationship with him and his commitment to us.

But what is the point of bringing our petitions to God in prayer? After all, the Bible says that "God knows what we need before you ask him" (Matt 6:8). (Note: this claim holds true regardless of one's view of God's relationship with time.) The Bible does not address this question; the practice of prayer is assumed (e.g., Matt 6:5–15; Luke 11:1–4). The gospel writers inform us that Jesus prayed regularly (Mark 1:35; Matt 14:23; Luke 5:16), so modeling the life of prayer for his followers. Nevertheless, what is the need of praying to an omniscient God, who knows all that can be known? Aquinas said: "We must pray, not in order to inform God of our needs and desires, but in order to remind ourselves that in these matters we need divine assistance."[46] Elsewhere Aquinas states:

> Prayer is not offered to God in order to change God's mind, but in order to excite confidence in us. Such confidence is fostered

---

44. For a pastoral introduction written by a scientist-theologian, see Wilkinson, *When I Pray What Does God Do?*

45. MacKay, *Science, Chance, and Providence*, 55.

46. Cited in Tiessen, *Providence and Prayer*, 196. Furthermore, Augustine comments: "God does not need to have our will made known to him—he cannot but know it—but he wishes our desire to be exercised in prayer that we may be able to receive what he is preparing to give" (ibid.).

principally by considering God's charity toward us whereby he wills our good.[47]

Prayer is therefore for *our* benefit, *not* God's, and—at best—results in us having a new perspective. This viewpoint is in-keeping with a classical view of God who transcends time and is immutable, that is, unchanging and unaffected by our prayers. Prayer in this framework can *only* change *us*, not God. Nor—for that matter—the created order, since the future is already determined in the mind of God. If we believe that the future is already so predestined then prayer cannot influence what God has already decided. If this is the case, in what coherent sense can we honestly say that God "responds" to our prayers? It is an important issue, particularly in the face of suffering. I am not alone in thinking such a view of God's providence is uninspiring. In 233 CE, Ambrose wrote to Origen:

> First, if God foreknows what will come to be and if it must happen, then prayer is in vain. Second, if everything happens according to God's will and if what He wills is fixed and no one of the things He wills can be changed, then prayer is in vain.[48]

We are left praying simply out of obedience, or because we believe we *should* pray; some may even feel guilty if they don't pray, but in their heart of hearts they don't really believe that their prayers are going to be effective or change the outcome.

Theologians have long wrestled with this issue. We saw in the previous chapter differing views on God's relationship with time, and these naturally lead to other models of God's providence.[49] Interestingly, dialogue between science and theology has resulted in helpful complementary insights. Polkinghorne gives two criteria for theological coherence in prayer:

> Prayer only makes sense within *a certain type of universe*. The mechanical world of Laplace's calculator where both past and future are inexorably contained within the dynamical circumstances of the present, would be too rigid a world to have prayer within it. . . . Prayer also makes sense only *with a certain kind of God*. A God totally above the temporal process, with the future as clearly

---

47. Ibid.
48. Cited in Sanders, *God Who Risks*, 277.
49. These are reviewed in Tiessen, *Providence and Prayer*.

present to him as the past, would be a suspect collaborator in the encounter of prayer.[50]

With these two in place, Polkinghorne concludes, prayer is not a "nonsensical idea" but becomes a "rational possibility."[51] It is only from such a position will we have the confidence to engage ourselves in the discipline of prayer. Prayer as a potential means of genuine change to the physical world—and not merely of psychological benefit—is totally undermined by a rigid clockwork universe. I suggest that legacy lives on in our subconscious, despite the developments of science over the last century. The mechanistic view of the world is dead, let us not resurrect it within our *theology* and so inhibit our view of God's capabilities and activities in the world.

Nevertheless, prayer is not magic and cannot change the facts of the present situation, just like the past cannot be altered. Too often our prayers treat God as if he was the master Magician—or they are a vain attempt to manipulate God. C. S. Lewis reminds us that prayer is a *request*:

> Now even if all the things that people prayed for happened, which they do not, this would not prove what Christians mean by the efficacy of prayer. For prayer is a request. The essence of a request, as distinct from compulsion, is that it may or may not be granted.... Invariable "success" in prayer would not prove the Christian doctrine at all. It would prove something much more like magic—a power in certain human beings to control, or compel, the course of nature.... Prayer is not a machine. It is not magic.[52]

Prayer is not formulaic, and neither can prayer's effectiveness be proved or disproved logically. Just because the request was "granted" does not mean that it would *not* have been granted had you *not* prayed. We are bound by the arrow of time; we cannot go back and run through the exact same scenario again, this time without prayer, and see if the same result is achieved. You need not, unless you choose, believe in the causal connection between

---

50. Polkinghorne, *Science and Providence*, 84, emphasis mine.
51. Ibid.
52. Lewis, "Efficacy of Prayer," 4–5, 9. MacKay writes: "From the biblical standpoint, the petitioner's prayer is not a matter of pulling invisible causal strings to bring about a desired answer. This is what chiefly distinguishes it from pagan superstition, where the deity (or lucky charm, or whatever) is envisaged as a source of intangible *forces* acting *within* the world by invisible causal connections that overpower the 'forces of nature.'" MacKay, *Science, Chance, and Providence*, 65.

the prayer and the result. The effectiveness of prayer, like the existence and significance of miracles, is a matter of faith.[53]

For others, prayer is unnecessary because there is a fatalistic expectation that God will always do what is "best" anyway. The myriad of complexities in an *open* world means that it is far from likely that there is only one "best action" for God. Rather there will be a range of creative alternatives open to God. Consequently, what is "best" if we do *not* pray might well be *different* from what is "best" if we *do* pray.[54]

Returning to the first question, why articulate prayer if God already knows what we want and need? Keith Ward points out that "God may know what we *want*, better than we do. But he only knows what we *request*, if we *actually* request it."[55] There is a difference between wishing and asking. We can wish for something without putting any conscious or physical effort to bring that desire about. In contrast, to *request* something of God requires us to *think* of him, *rely* on his ability, and *trust* in him. It is both an act of our will and faith. This is why it is necessary for us to articulate our request in prayer, and not just hope that he might give us what we desire.[56]

How God responds to our requests we cannot say, since we do not know the constraints of the whole system or the involvement of others—not forgetting that they also have free will. Nevertheless in the complex web of possibilities, our prayer may enable God to shift the constraints in a favorable direction to respond to our prayer. Making the request itself may *shift* the balance. In this way our prayer can have a positive contribution (i.e., become a causal factor) in the realization of God's purposes in the world.[57] Divine action involves *our* cooperation.

In summary, Polkinghorne writes:

> Prayer is neither the manipulation of God nor just the illumination of our perception, but it is the alignment of our wills with his, the correlation of human desire and divine purpose. That alignment is not just the passive acceptance of God's will by human resignation (though "if it be thy will" is an essential part of any

---

53. Polkinghorne surmises: "Personal experience is irreducibly individual, and in consequence its record is inescapably anecdotal." Polkinghorne, *Science and Providence*, 85.

54. Ward, *Divine Action*, 161–62.

55. Ibid., 162, emphasis mine.

56. Ibid., 163.

57. Ibid. I would add, either by praying aloud or by formulating a deliberate mental or inner prayer to God.

prayer, since God is the necessary partner in it), but it is also a resolute determination to share in the accomplishment of that will (so that prayer is never divorced from action, nor a substitute for it). Prayer is a collaborative personal encounter between man and God, to which both contribute.[58]

Theologian Clark Pinnock, coming from a very different starting point, arrives at the same conclusion:

> God could act alone in ruling the world but wants to work in consultation. It is not his way unilaterally to decide everything. He treats us as partners in a two-way conversation and wants our input. . . . He enlists our input because he wants it, not because he needs it. He treats us as responsible agents with whom he has a dynamic relationship. . . . God does not stand at a distance but gets involved, becomes conditioned, responds, relents, intervenes and acts in time. Prayer changes things because God allows it to influence him so that prayer becomes an effective contributor to the flow of events.[59]

Since our prayers become part of the complex causal matrix, prayer will *always* make a difference to the world—even if it does not expressly give us the outcome we desire. This foundation should inspire us to pray.

---

58. Polkinghorne, *Science and Providence*, 81.

59. Pinnock, *Most Moved Mover*, 171–72. Pinnock cites Richard J. Foster, author of *Celebration of Discipline*, as endorsing his view, 173–74. Furthermore, Dallas Willard writes: "God's 'response' to our prayers is not a charade. He does not pretend that he is answering our prayer when he is only doing what he was going to do anyway. Our requests really do make a difference in what God does or does not do." Willard, *Divine Conspiracy*, 244.

Chapter 8

Revisiting Science and Scripture
Creation Texts in the Old Testament

> Praise him, sun and moon; praise him, all you shining stars! Praise him, you highest heavens, and you waters above the heavens! Let them praise the name of the LORD, for he commanded and they were created. He established them forever and ever; he fixed their bounds, which cannot be passed. Praise the LORD from the earth, you sea monsters and all deeps, fire and hail, snow and frost, stormy wind fulfilling his command! Mountains and all hills, fruit trees and all cedars! Wild animals and all cattle, creeping things and flying birds! Praise the LORD. —Psalm 148:3–10, 14b

INTRODUCTION

Moving beyond the tensions that lead to conflict between science and Christianity, how are we to read biblical texts relating to creation in light of modern science? Is it *possible* to harmonize early Genesis with the findings of science? Should we even try? What are we to make of other creation texts, such as Job 38–41? That is the theme of this final chapter, giving Scripture the last word.

David Wilkinson comments briefly on various methods employed over the last two centuries that attempt to relate the Bible and science.[1]

---

1. Wilkinson, "Genesis in Light of Modern Science," 127–44.

Regarding Genesis 1, one popular approach is to introduce the six days of creation as ages or epochs of time. Consequently the Hebrew word for "day" (*yom*) is interpreted figuratively in terms of an unspecified period of time which is then linked to the millions of years required for the evolutionary process and the fossil record. However, this is problematic since it disrupts the author's rhythmic use of "evening and morning" (1:5, 8, 13, 19, 23, 31) on each of the six days of God's creative acts.[2] It seems to me that it is *exegetically* unacceptable to interpret the text in this way. Rather, this approach is an example of *eisegesis* where one reads *into* the text the desire to see harmony—or concord—with the timescales required by geology and biology.[3] We need to move beyond seeing Genesis 1–3 as a divinely inspired explanation of origins in a *scientific* sense (or a *historical* one). To claim that does not mean that the Israelites were not interested in knowledge about nature. We are told in 1 Kings 4:33 that Solomon's wisdom included insights on the plant and animal kingdoms. However that knowledge was *prescientific* and this is evident in God's use of the "dry land" and "waters" to mediate creation (Gen 1:2, 11, 20, 24). All this liberates us from attempts to shape Scripture in order to comply with the findings of science, and vice versa, and allows us to have a more enriching conversation.[4] In such a dialogue we are free to explore many things, including the issues of origins and environmental responsibility, and to address the question: "What does it

---

2. Moreover Exod 20:11 and 31:17 only make sense if the "days" are actual days.

3. See also the discussion of concordism in chapter 4.

4. Walter Brueggemann writes: "At the outset, we must see that this text (Gen 1:1—2:4) is not a scientific description, but a theological affirmation. It makes a faith statement.... This text has been caught in the unfortunate battle of 'modernism', so that the 'literalists' and 'rationalists' ... [are] nearly ready to have the text destroyed in order to control it. Our exposition must reject both such views.... Rather, it makes the theological claim that a word has been spoken which transforms reality.... The claim made is not an historical claim but a theological one about the character of God who is bound to his world and about the world which is bound to God.... In interpreting this text, the listening community must speak its own language of confession and praise, which is not the language of 'scientific history' nor the language of 'mythology and rationalism'. These tempting epistemologies reflect modern controversies and attest to a closed universe... . Against both, our exposition must recognize that what we have in the text is *proclamation*. The poem does not narrate 'how it happened' ... [rather] Israel is concerned with *God's lordly intent*, not his *technique*. Conversely, the text does not present us with what has always been and will always be: an unchanging structure of world. Rather, the text proclaimed a newness which places the world in a situation which did not previously exist.... Our interpretation must reject the seductions of literalism and rationalism to hear the news announced to the exiles. The good news is that life in God's well-ordered world can be joyous and grateful response." Brueggemann, *Genesis*, 25–26, his emphasis.

mean to be human?" Our responses to those matters have important ethical consequences and better inform us about the God we worship. In this chapter I will briefly explore some of the biblical texts that relate to creation. This is not meant to be an exhaustive exegetical or theological study, or a cultural analysis of the texts; that is beyond my capabilities and there are many good sources on those topics.[5] Rather just to highlight some perennial features that arise when studying Scripture from a modern viewpoint informed by science.

GENESIS 1

It has often been pointed out that the opening chapters of Genesis contain two separate creation stories that have been carefully combined by later redactors. The first account (Gen 1:1—2:4a) is assigned to a priestly writer, and the second (Gen 2:4b–25) to an author who knows God as YHWH, rather than *Elohim*. It also is important to remember that the Genesis that we have today emerged in its final form at the time of the exile (sixth century BCE). As such there is a contrast between the God of Israel and the Babylonian deities. Consequently, these two creation narratives need to be appreciated in the context of the stories of origins from the neighboring cultures of Mesopotamia, Canaan, and Egypt. All this goes to show is that the writers and redactors were telling Israel's *own* story in a given *context*, rather than some universal narrative articulated in an abstract manner for the whole of humankind. These early chapters of Genesis describe Israel's own understandings of themselves and, at a time of dispersion and exile, they become community-defining texts that affirm their God-given identity—one that is covenantal (Adam, Noah, Abraham, Moses, David) from the very beginning.

In Genesis 1, we see that *God* is the primary subject of this chapter and whose ultimate origin is unquestioned by this community of faith. The liturgy-like poetry introduces a seven-day structure ending with a Sabbath, most appropriate if the writer is of a priestly class. While seven is the number of completeness, unity, and perfection, eight creative acts are to be found within six days (two acts occur on days 3 and 6). Rather than viewing God's activities on these six days in a literal sense, or one that is meant to correspond to a scientific sequence, it is better to view the days in a *literary* fashion. Table 1 provides a framework of God's activity in Genesis 1, in

---

5. E.g., see Fretheim, *Relational Theology of Creation*.

which God first *separates* spaces or regions (days 1–3) and then *fills* each of those spaces (days 4–6). This elegant schema is not too rigid, resulting in the text being artificially constrained; rather it mirrors a literary *pattern* corresponding to the general theme of God bringing *order* out of disorder (1:2).[6]

Table 1: A Framework of God's Creative Activity in Genesis 1

God *Creates* "Spaces" (or Domains)	God *Fills* "Spaces" (or Domains)
Day 1	Day 4
God separates light from darkness (v4).[A]	God fills the sky with lights: the sun, moon and stars—to rule the seasons and maintain the separation between darkness and light (v14–18).
Day 2	Day 5
God separates the sky from the "waters" (v7, 8).	God fills the waters with living creatures[B] and the sky with birds (v20–22).
Day 3	Day 6
(a) God separates the land from the seas (v9, 10). (b) God fills the land with *vegetation*.	(a) God fills the land with domestic and wild *animals* (v24, 25). (b) God makes humankind (v26–30).

A. Fretheim points out that light was thought to have another source (Job 38:19; Isa 30:26) and only augmented by the sun. (e.g., light on a cloudy day, and before sunrise and after sunset.) Fretheim, "Genesis," 343.

B. God also made fear-inducing sea monsters (1:21) that were often, particularly in neighboring cultures, associated with chaos; this matter will be addressed later.

There is a poetic regularity to each day's activities:[7]

1. Command: "God said let there be . . . "

2. Execution: "And it was so."

3. Assessment: "God saw it was good."

4. Sequence/Time: "There was evening and morning . . . "

While this pattern is not perfectly symmetrical throughout all the six days, the overall effect is to give a melodic crescendo that peaks at the end of

---

6. On day 3, God not only creates the *space* of dry land but provides vegetation of all kinds to make it *habitable*, or ready, for all animal life and humankind who will fill the space on day 6. In light of Gen 1:28–29; 9:2–3, all air-breathing animals were intended to be vegetarian! In keeping with this picture, Isaiah 11:7; 65:25 imply that animals will be herbivorous in the new creation.

7. Brueggemann, *Genesis*, 30.

day 6, followed—appropriately—by relaxation and blessing on the seventh (Sabbath) day. Indeed, as theologians remind us, the true climax is on the seventh day with the story beginning and ending with God—not culminating with the creation of humankind! We see in each of the six days that God's speech is actionable and nothing thwarts God's intentions. Creation is not an accident but a deliberate act of the divine will.[8] Moreover, God approves and delights in his creation, affirming it as "very good" at the end of day 6. Nevertheless, "good" does not imply a static state of perfection or a creation in no need of further development.[9] Rather, creation is purposeful and dynamic; the potential of becoming is built into the very structure of things. Furthermore, for creation to be called "good" means that nothing God has created is inherently evil.[10]

In addition, just as God delegates the sun and moon "to rule" the days and seasons (1:16), so God delegates humankind "to rule" over the fish, birds and every living creature (1:28).[11] God is, evidently, a power-sharing God. The command to "be fruitful and multiply" (1:22, 28) is given to the creatures of the sea, the birds of the air, and to humankind (but, oddly, not to land-based creatures). This command, which follows God's blessing, gives permission to creatures to *be* the "other" and further demonstrates God sharing his creative capabilities. God's blessing, in this context, is a word of empowerment.[12] This is made even more special in the case of creating humankind, since God consults the divine council ("let *us* make humankind in *our* image," 1:26).[13]

In light of the industrial and technological revolutions, and the resulting environmental crisis, some have placed responsibility for this present calamity at the feet of the church. As discussed in chapter 4, it has been

8. See Fretheim, "Genesis," 343.

9. See ibid.

10. See Fretheim, *Relational Theology of Creation*, 52, 56.

11. Fretheim writes: "The fact that the sun and moon are not specifically named [in 1:16], and the stars are just mentioned, may reflect a polemic against the religious practices in Mesopotamia, where heavenly bodies were considered divine and astrology played an important role in daily life. All are here acclaimed as the creations of the one God." Fretheim, "Genesis," 344.

12. Fretheim, *Relational Theology of Creation*, 50.

13. See also Gen 3:22. The notion of a heavenly court is found elsewhere in the Old Testament (e.g., Job 38:7; 1 Kgs 22:19; Isa 6:8; Jer 23:18). In light of John 1:1–3, some Christians interpret this plurality in a *Trinitarian* way. While this is understandable, we must nevertheless recognize that this is eisegesis—an example of an imposition of later *Christian* theology on the *Jewish* Scriptures.

claimed that since modern science emerged within Christendom, the quest for control *over* nature arises because it is mandated in Genesis 1:28—with its command to "subdue" and "have dominion"—giving divine license to rape the land, sea, and sky. This accusation is, in my view, over-stated, given science's rejection of its monotheistic roots over the last century or more. Nevertheless, this does not let the church off the hook for being responsible caretakers of God's good creation (see also Gen 2:15). Whatever else "being made in the image (or likeness) of God" may imply, Bernhard Anderson states that "Adam is created to be God's *representative* on earth."[14] He continues:

> Viewed in this perspective, Adam is not an autonomous being, at liberty to rule the earth arbitrarily or violently. On the contrary, human dominion is to be exercised wisely and benevolently so that God's dominion over the earth may be manifest in care for the earth and in the exercise of justice.[15]

Commenting on 1:28, Terence Fretheim states:

> A study of the verb "have dominion" reveals that it must be understood in terms of caregiving, even nurturing, not exploitation. As the image of God, human beings should relate to the nonhuman as God relates to them. . . .The command "to subdue the earth" focuses on the earth, particularly cultivation (see 2:5, 15), a difficult task in those days.[16]

Moreover, Walter Brueggemann reminds us that the "dominion" that is mandated toward animals is like that of a shepherd, who cares for and feeds their flock. It does not legitimize abuse or exploitation. "It has to do with securing the well-being of every other creature and bringing the promise of each to full fruition."[17] The exegetical and theological cases are overwhelm-

---

14. Anderson, *Contours of Old Testament Theology*, 90.

15. Ibid., 91.

16. Fretheim, "Genesis," 346, emphasis mine. Note that the Hebrew word for "earth" is also used for "soil," "dirt," and "ground"—but *not* the modern notion of a planet! Fretheim adds: "This process offers to the human being the task of intra-creational development, bringing the world along to its fullest possible creational potential. Here paradise is not a state of perfection, not a static state of affairs. Humans live in a highly dynamic situation. The future remains open to a number of possibilities in which creaturely activity will prove crucial for the development of the world."

17. Brueggemann, *Genesis*, 32. He also points to Ezek 34 with its analogy of a leader as a shepherd and the prophetic warning to those who have misused the imperative of the Creator.

ing; as God's agents, humankind *is* divinely mandated to be responsible for creation. Christians—Christ's ambassadors—must, therefore, advocate for environmental concerns, animal welfare, and the ethical use of genetic manipulation.

The creation week is completed with the seventh day where God rested after finishing all the work that he had done (2:2–3). By blessing that day, the priestly author is claiming that the Sabbath is instigated by God, not humankind. Fretheim concludes:

> "Finishing" does not mean that God will not engage in further creative acts. . . . These days did not exhaust the divine creativity! The seventh day refers to a specific day and not to an open future. Continuing creative work will be needed, but there is a "rounding off" of the created order at this point.[18]

GENESIS 2–3

The second account of creation (Gen 2:4b–25) has a very different tone from that of the first (Gen 1:1—2:4a). The lofty elegance of the first story often has more appeal in contrast to the relative simplicity, even naivety, of the second (and probably much older) narrative. Yet the redactors carefully blended both stories together as part of the prehistory of Genesis 1–11 and it is unwise to pull the canon of Scripture apart. In 2:7 we are told that God formed man from the dust of the ground. The use of dust or clay (see Job 10:8–9) presents God as a potter, a common metaphor within the Old Testament.[19] Once the clay man was formed ("*Adam*" has Hebrew wordplay with the *adama*, meaning "ground"), he became animated when God's breath of life was breathed into him.[20] God is also presented as a farmer or gardener (2:8) who planted the fertile garden of Eden to provide for all of Adam's needs.[21] Two mysterious trees are introduced: the "Tree of Life" and

---

18. Fretheim, "Genesis," 346.

19. See, e.g., Isa 41:25; 45:9; 64:8; Jer 18:1–6; Sir 33:10–13.

20. Gen 7:21–22 makes a connection with God's "breath of life" animating all land animals, so making them living beings.

21. Fretheim notes there are "some twenty images of God" discernable in Gen 1–2 that are correlated to modes of creation, including Maker, Speaker, Potter, Builder, Surgeon, Architect, and Evaluator. See Fretheim, *Relational Theology of Creation*, 36–48. Incidentally, "paradise" is a Persian loanword in Hebrew, Aramaic, Syriac, and Greek that signifies a beautiful enclosed garden, that of a king. (See Charlesworth, "Paradise," 377.)

the "Tree of the Knowledge of Good and Evil" (2:9). The Tree of Life has its historical roots in a royal image of a task given by the gods to a king to guard and nurture the mystery of life; there is very little known about the Tree of Knowledge—other than its mention in Genesis 2–3.[22] We are not told why they are there; perhaps this is storytelling at its best—introducing ominous elements that leave the reader in suspense! Later in the story there is permission or freedom (2:16) and prohibition or restriction (2:17). Any good audience to the oral story will know that such a prohibition is a prediction of a plot line that will challenge that rule!

Moving on to Genesis 3, the serpent is described as "wise" (Gen 3:1), which is sometimes regarded as "crafty" or "shrewd," perhaps earthly—rather than Godly—wisdom (e.g., 2 Cor 11:3—"cunning").[23] We have similar associations today: e.g., owls are wise, foxes are cunning or sly. Ironically, this characteristic of a snake is also referred to as something to be emulated: "be wise as serpents" (Matt 10:16). Another feature of the serpent is that it talks, which in ancient literature signifies a supernatural presence—not as a snake with extraordinary vocal cords.[24] Furthermore, a serpent often has negative, dark, or evil associations. In Egypt, Apophis, often depicted as a serpent, is a god of chaos who threatens the very stability of the cosmos and is opposed to the sun god. There is also a link with Tiamat who is often portrayed in Mesopotamian art as a serpent or dragon. In the Gilgamesh Epic, Gilgamesh finds a plant which has the power to give him everlasting

---

The concept evolved to imply crystal clear water, trees and flowers in constant bloom, a place with no sickness and at an ideal temperature for human habitation. Interestingly, the kings in Mesopotamia often described themselves as great gardeners. The garden of Eden of Gen 2–3 is also referred to elsewhere in the Old Testament (e.g., Isa 51:3, Ezek 31:9, 16, 18; and Joel 2:3); this land is associated with an abundance of fertility and where death and sterility are absent. Anderson, "Eden," 186–87.

22. Brueggemann, *Genesis*, 45.

23. In Gen 3, the Hebrew word for the snake's wisdom (*arum*) appears to be a word play with *arummim*, meaning "naked"; this subtly is, of course, lost in translation. This might suggest that humankind is vulnerable to being exposed to nature's charms. What the serpent is *not*, at least at this stage in Old Testament literature, is a euphemism for the satan, the "adversary" or "accuser." (See Conrad, "Satan," 116.) This connection is made later, e.g., Wis 2:24; Rev 12:9; 20:2. In the same way the sin of Adam is rarely discussed (or commented on) by later Old Testament authors, although the pervasiveness of sin in assumed in 1 Kgs 8:46; Jer 13:23; and Ps 51.

24. See the other scriptural reference of a talking animal: Balaam's donkey, Num 22:22–35. Incidentally, in the Egyptian tale *The Shipwrecked Sailor* a man on a mystical island encounters a talking snake—with a beard and a gold skin, as they were associated with a king/deity. See Hodge, *Revisiting Genesis*, 113.

life, but it is stolen from him by a serpent.[25] These negative associations would then warn the *reader* that the serpent in Genesis 3 is dangerous, even opposed to God, something of which the first humans were innocently unaware. That being the case, it should also be noted that (in this story) the serpent does not challenge God directly, but indirectly through the people God has made.

There is also a common understanding in the ancient world which links the uncultivated fields, wilderness, and desert lands with "death" (or death's *realm*).[26] In this region the wild beasts rule and, in Egyptian imagery, Apophis (the serpent) dwells. In Genesis 3 the difference is that the serpent is, for some reason, *inside* the fertile garden—even though in Genesis 3:1, the snake is regarded as a "wild" animal (rather than a domesticated beast of burden—Genesis 3:14—or some form of deity, or explicitly evil). This differentiation of the kinds of animals is not a property of the animals themselves, but "mark of the boundaries between the civilized and uncivilized lands. These then become symbols of creation and chaos, and as such represent the land of the living and the land of the dead."[27] It is important to note that since the author identifies the snake as a wild animal, it is a feature of *creation* and so *external* to both Adam and Eve; i.e., the origin of the "temptation" is not deemed to be an *internal* conflict.

Genesis 3:22 confirms the serpent was actually telling the truth on (at least) one matter: "God knows that when you eat of it your eyes will be opened, and you will be like God, knowing good and evil" (Gen 3:5). After they ate the fruit, God's concern moved toward their access to the Tree of Life and one aspect of the story's culmination is to comment on humanity's search for immortality and "explain" why it is now totally unobtainable (i.e., the cherubim, Gen 3:24). After eating the fruit of the Tree of Knowledge, God acknowledges that they *could* still live forever if they were able to eat the fruit of the Tree of Life. Fretheim writes: "The expulsion does not mean innate immortality has been lost; rather the possibility of ever attaining it has been eliminated."[28] The expulsion from the garden—a place created,

---

25. Ibid., 115. This "explains" why snakes can shed their skins and appear to have new life or become young in old age. The connection of serpents with immortality (not just wisdom and chaos) needs to at least be recognized in the Genesis 3 account, since the end of the story points to potential immortality as the reason for their expulsion from the garden.

26. Ibid., 117.

27. Ibid., 118.

28. Fretheim, "Genesis," 364. Elsewhere, Fretheim writes: "If they were created

organized, ruled by God, and conducive to life—places Adam and Eve in a hostile wilderness that threatens human existence.[29]

Most Christians understand this chapter as describing "the Fall," implying a fall *down* from a state of moral *and* physical perfection. However, if Adam and Eve were "perfect," how could they have failed? "Rather they were 'good,' which entails considerable room for growth and the development of potentialities."[30] Some see the Fall as falling *upward*, with the maturation of humankind's consciousness and the development of moral responsibility. A difficulty with this view is that the prohibition in 2:17 does not portray God in a positive light, but as a parent who opposes maturity (see also 3:22). The incident makes God appear to be threatened by the prospect of human's gaining knowledge, and to be overreacting when humans transgress his arbitrary limits.[31] Alternatively, the falling "upward" movement emphasizes the desire for a creature to become like their Creator and determine ("know") right from wrong (3:5). As human beings strive upward for godlike powers they fall down as they are not able to handle the consequences of their own decisions. Another way to view the "Fall" metaphor is to see it as a falling *out*. Whatever else, the story is certainly about the breakdown of *trust* and of *relationship*; between people and God, between Adam and Eve (and, later, Cain and Abel), and between themselves and the created order.[32] By not *trusting* the Creator they fall out of relationship with him, resulting in separation, alienation, and disharmony.[33] The Hebrew concept of *shalom* is a restoration of peace with God, with each other, and with the created order. This is a reversal of the falling out of relationship and, for the Christian, *shalom* is ultimately achieved through the birth-life-death-resurrection-ascension of Jesus Christ.

---

immortal, the tree of life would have been irrelevant. Death *per se* was a natural part of God's created world." Fretheim, "Is Genesis 3 a Fall Story?," 152.

29. Hodge concludes, "The serpent, as a symbol of wisdom, makes no sense in this context if not combined with the serpent the symbol of chaos as well." Hodge, *Revisiting Genesis*, 119.

30. Fretheim, "Genesis," 368.

31. Fretheim, *Relational Theology of Creation*, 71.

32. Ibid., 74.

33. Note too that the final form of the text emerged in the context of the Babylonian exile. This is also alienation and an enforced expulsion from their homeland, one brought about—as they understood it—by their persistent breaking of their covenantal *relationship* with God.

In the final analysis, we need to be mindful to not force *our* modern questions onto the narrative or to overanalyze it by scrutinizing the holes in the plotline.[34] Since the story was for the benefit of its original hearers, we must assume it was sufficient for its intended purpose.[35] This story does not, however, nor does the Bible as a whole, seek to explain the *origin* of evil but to "witness to its character as guilt and as the unending burden that humankind bears."[36]

But what about physical death? Christians, using Romans 5:12–21, read the cause of physical death back in to Genesis 3 story. Paul's complex theological argument juxtaposes Adam and Christ and assumes the *reality of death* and its connection to Genesis 3 via intertestamental wisdom literature.[37] Paul *interpreted* Jewish Scripture in the context of his time—as did the gospel writers—but his main theological point emphasized what God has done in Jesus.[38] But we should not infer from this that prior to Genesis 3 there was no cycle of birth-life-death-decay.[39] In addition to the fossil record, we have, after all, oil in the ground whose origin requires the death of

---

34. Walter Brueggemann emphasizes that (a) the biblical authors (even Paul) are not as concerned with this passage as, say, Augustine and the birth of the notion of "original" sin, (b) the text does not seek to explain how evil came into the world, (c) neither is it an account of the origin of death (a mechanistic link between sin and death), and (d) neither is it a narrative about the evils brought about by sex! See Brueggemann, *Genesis*, 42–43.

35. For an example of a modern retelling of the Adam and Eve story, see Reddish, "Dawn," 15–26.

36. Bonhoeffer, *Creation and Fall*, 105.

37. Sir 25:24; Wis 2:24; 2 Esd 3:7–22; 7:118.

38. This is discussed at length in Enns, *Evolution of Adam*, 79–135.

39. Polkinghorne writes: "[Adam and Eve's] turning from God did not bring biological death into the world, for that had been there for many millions of years before there were any hominids. What it did bring was what we may call 'mortality,' human sadness and bitterness at the inevitability of death and decay. Because our ancestors had become self-conscious, they knew long beforehand that they were going to die. Because they had alienated themselves from the God whose steadfast faithfulness is the only (and sufficient) true ground for the hope of a destiny beyond death, this realization brought deep sorrow at the transience of human life. . . . Alienation from God brought the bitterness of mortality, but the relationship of humanity to God has been restored in the atonement (at-one-ment) brought by Jesus Christ, in whom the life of humanity and the life of divinity are both present and the broken link is mended." Polkinghorne, *Testing Scripture*, 30; see also Polkinghorne, *Reason and Reality*, 99–104. Paul, in 1 Cor 15, again uses the Adam-Christ contrast and articulates the great Christian hope of ultimate resurrection and restoration. There Paul also writes of the *sting* of death being swallowed in victory; Christ's resurrection (signifying the first fruit of the final harvest) removes the fear from our comprehension of our own mortality introduced in the Fall narrative.

organic vegetation. Those who claim that death and decay only arose after Genesis 3 often also maintain that the laws of nature changed at this time. In other words, there was a real ontological change in God's created order such that—some claim—creation is no longer "good." However, most theologians emphatically reject such a claim, and there is no such discontinuity in the scientific record. Moreover, if there were no yearly cycle of spring and autumn, and *if* you could have chopped down a tree in the garden of Eden, it would not have displayed annual growth rings! Either the cosmic and geological timescales are an elaborate lie, or we need to find a different way to understand this story.

No Christian doubts the *theological* significance of the Genesis 3 story but there is no need claim historicity to the events and thereby make it an unnecessary hurdle for faith.[40] The same is true with the story of Noah. In the flood narrative we read of God's intent to purge the world of its corruption. God wanted to undo creation and to begin again (6:7). The waters that God separated to form the sky and the land on days 2 and 3 in Genesis 1 are allowed to wreak havoc (7:11–12), reversing the movement toward chaos for a time (8:2). Yet, the old world was not completely destroyed (8:11) and the recognizable preflood creation emerges as the waters subside. Of particular importance is God's universal covenant with Noah (Gen 8:21—9:17). Here we hear the blessing of God from Genesis 1 being *reiterated* (8:17; 9:1, 7) and the reaffirmation that despite all that has happened (the "Fall," murder, etc.) humankind is *still* made in the image of God (9:6) and God remains committed to the world he made.[41]

CREATION AND CHAOS IN OLD TESTAMENT WISDOM LITERATURE

In this section the relationship between creation and chaos, as embodied in biblical references to the sea and its monsters, will be explored.[42] Mention of the mysterious sea monsters (Leviathan and Rahab) seem bizarre and are largely ignored by most Christians. How are we to understand such texts? And do they have anything to say to contemporary Christians? I think they do, since they represent a complementary depiction of creation from that

---

40. For further discussion, see Brueggemann, *Genesis*, 41–44, 53–54.

41. For further discussion, see, e.g., Fretheim, "Genesis," 388–402, and Brueggemann, *Genesis*, 73–88.

42. Hiebert, "Chaos," 582.

of Genesis 1–2—one that is often overlooked. Such references indicate that untamed chaos has a God-ordained place within creation.

Jon Levenson sees two different forms of chaos in the Old Testament: (a) inert matter lacking order and so requiring differentiation (e.g., potter and clay metaphor, Gen 2:7–9) and (b) chaos as a living being with its own will and personality that is at cross purposes with God and must be vanquished before God can create the cosmos. This borrowed imagery comes from the creation myths of Israel's neighbors.[43] Genesis 1 can be understood in the context of (a). However there are a number of creation references within the Wisdom Tradition (e.g., Psalms, Proverbs, Ecclesiastes, and Job) that are articulated in terms of (b).

The stories of origins of Israel's neighboring cultures usually have the world as being formed by a mythic battle against water (sea, ocean) that signifies "chaos" and are often personified by a chaos dragon.[44] For example, in the Ugaritic text of the Baal/Yam battle, Yam (sea monster) claims kingship of the world, and all the gods accept Yam's superiority except Baal (storm god), and a battle ensues. Baal subdues Yam and either kills the dragon or confines it to its proper place.[45] This story institutes Baal's kingship over the earth. The Babylonian *Enuma Elish* account also has a preexistent salt-water chaos, Tiamat, that Marduk must defeat before creation can take place. Tiamat's corpse is cut in two and provides the material from which the sky and earth are made. Marduk uses the blood of another slain god (Kingu) to create humankind to serve the gods.

Certain Old Testament texts, discussed below, are clearly influenced by those stories of victory over the primordial sea. Indeed, it is hard to conceive that the author (or final redactor) of Genesis 1–2 would be unaware of those Babylonian and Canaanite myths.[46] Nevertheless, there is a significant difference from these foreign stories, in that no conquest is required by God before—or indeed, as part of—God's creative acts. Furthermore, there is a differentiation between the manipulation of the "formless void," "face of the deep/waters" (Gen 1:2), and the *separation* of the waters (Gen 1:6–7),

---

43. Levenson, *Creation and the Persistence of Evil*, 3–52. McGrath makes the same point: McGrath *Christian Theology*, 217.

44. Wright, "Cosmogony, Cosmology," 755–63; Van Der Toorn, "Baal," 367–69; Boyd, *God at War*, 75–79.

45. Day, "God and Leviathan," 425–29.

46. Fretheim, "Genesis," 356.

and the *creation* of sea monsters (Gen 1:21).[47] Consequently, the cosmic monotheism of Genesis 1 is to be contrasted to that within the *Enuma Elish* (or the Baal-Yam myth), despite the similar imagery of primordial "water."

God's mastery over the sea and its monsters is also evident in the wisdom literature, but the use of this imagery is not uniform. There are at least four different ways in which the sea and its monsters are depicted. One usage is that God simply *confines* the sea (i.e., no sea monsters are mentioned, or the waters are not personified), which is illustrated below (emphasis added):

> . . . when *he* [the LORD] *assigned to the sea its limit, so that the waters might not transgress his command*, when he marked out the foundations of the earth. . . . (Prov 8:29)

> When the Lord created his works from the beginning, and, in making them, *determined their boundaries*, he arranged his works in an eternal order, and their dominion for all generations. (Sir 16:26–27)

The psalms also make multiple references to the sea, for example:

> By the word of the LORD the heavens were made, and all their host by the breath of his mouth. *He gathered the waters of the sea as in a bottle; he put the deeps in storehouses.* Let all the earth fear the LORD; let all the inhabitants of the world stand in awe of him. For he spoke, and it came to be; he commanded, and it stood firm. (Ps 33:6–9)

While this passage refers to God's creative and restraining acts, there may also be an allusion to the Exodus (i.e., crossing the Red Sea), hence a literary merging of God's creative and redemptive acts.[48] This insinuation is clearer in Psalm 89:

> O LORD God of hosts, who is as mighty as you, O LORD? Your faithfulness surrounds you. *You rule the raging of the sea; when its waves rise, you still them. You crushed Rahab like a carcass; you scattered your enemies with your mighty arm.* (Ps 89:8–10)

---

47. See also 2 Esd 6:41, 47–52.
48. Clinton McCann, "Psalms," 810.

Although Rahab (meaning "boisterous or stormy one") can refer to a chaos dragon, it is also a mythical name for Egypt.[49] This connection is even clearer in Psalm 77:16–19 (and Isa 51:9–10).

The third way in which the stormy sea is used within the wisdom literature makes more graphic reference to the Tiamat and Yam imagery in the context of creation, for example:

> It was you who split open the sea by your power; you broke the heads of the monster in the waters. It was you who crushed the heads of Leviathan and gave him as food to the creatures of the desert. It was you who opened up springs and streams; you dried up the ever flowing rivers. The day is yours, and yours also the night; you established the sun and moon. (Ps 74:13–14)

In this text, like Psalm 89:8–10, God has explicitly killed the ferocious sea monster, Leviathan.

In the first usage, however, it is clear that God *does not eradicate* the seas (or waters) but allows them to *function within boundaries or limits*. This latter portrayal is also present in Job, where there is also often explicit reference to Leviathan/Rahab:

> He binds up the waters in his thick clouds, and the cloud is not torn open by them. . . . He has described a circle on the face of the waters, at the boundary between light and darkness. . . . *By his power he stilled the Sea; by his understanding he struck down Rahab. By his wind the heavens were made fair; his hand pierced the fleeing serpent.* (Job 26:8–13)

> Or *who shut in the sea with doors when it burst out* from the womb?—when I made the clouds its garment, and thick darkness its swaddling band, *and prescribed bounds for it, and set bars and doors, and said, "Thus far shall you come, and no farther, and here shall your proud waves be stopped"*? (Job 38:8–11)

While God's power is very evident in these texts, there is still a persistence to the presence of the sea. The sea may be confined, but it is not tamed. Janzen writes:

49. Ibid., 1035, and Mays, *Psalms*, 281, 284–85; see also Ps 65:7; 74:13–14; 87:4. This connection with Egypt is based on the premise that the enemy of God's covenant people is, by extension, an enemy of God. The use of this metaphor is a polemic reminder that the God of Israel is more powerful than any neighboring foe (or their deities). Care is needed, however, as polemics is a rhetorical device and so the linking of Israel's enemies with evil should not be understood literally.

> The Sea appears as a chaotic energy threatening destruction; and cosmic order with its life-giving and meaningful forms presupposes the effective limitation of this energy.... [In view of these allusions earlier in Job (i.e., 7:12; 9:8; 26:12; 28:14, 22)], it is unlikely that the sea is presented in 38:8–11 in demythologized fashion, that is, as simply a natural region. The overtones of primal chaos are unmistakable.... The sea, even as primal chaos, is limited to, yet given, a place in the scheme of things.[50]

And Newsom adds:

> The chaotic waters have a place in God's design of the cosmos, yet one that is clearly circumscribed. They are the object not only of divine restriction but also of divine care.[51]

In Psalm 104, Leviathan is not only part of creation but was also formed, or made, for "sport":

> O LORD, how manifold are your works! In wisdom you have made them all; the earth is full of your creatures. Yonder is the sea, great and wide, creeping things innumerable are there, living things both small and great. There go the ships, *and Leviathan that you formed to sport in it.* (Ps 104:24–26)

This "playful" or mocking portrayal is echoed or expanded (depending on dating) in Job 41:

> Can you draw out Leviathan with a fishhook, or press down its tongue with a cord? Can you put a rope in its nose, or pierce its jaw with a hook? Will it make many supplications to you? Will it speak soft words to you? Will it make a covenant with you to be taken as

---

50. Janzen, *Job*, 234-5. He adds: "What is remarkable about the present treatment of this standard theme [i.e., creation, primordial waters] is its ambivalence. On the one hand the Sea is restrained by bars and doors. On the other hand the birth of this same Sea is attended by God as by a midwife who carefully swaddles the infant in protective wrappings. This ambivalence, in which the Sea is surrounded by an action at once restraining and sustaining, is reflected in the structure of the passage.... All attempts to exegete the book of Job in such a way as to arrive at the conclusion that God there is indifferent to matters of justice overlook the fact that the place of the sea in the cosmos is delimited by divine decree. Perhaps the issue can be stated only in the modes of poetry. As for Job, he must come to terms with the brute fact of the place of the sea in the scheme of things; yet he is not to interpret it in such a way as to imply God's disinterest in law or justice." Ibid., 234–35.

51. Newsom, "Job," 602.

> your servant forever? Will you play with it as with a bird, or will you put it on leash for your girls? (Job 41:1–5)

Only God can confront these creatures (Job 40:19; 41:10–11) that no human (or other gods, Job 41:9, 25) can tame.

The fourth—and most rare—application is the eschatological reference to Leviathan in Isaiah 27:[52]

> On *that day* the LORD with his cruel and great and strong sword *will* punish Leviathan the fleeing serpent, Leviathan the twisting serpent, and he *will* kill the dragon that is in the sea. (Isa 27:1)

It should be noted that John the Seer also envisages a time when chaos will finally be defeated:

> Then I saw a new heaven and a new earth; for the first heaven and the first earth had passed away, *and the sea was no more*. (Rev 21:1)[53]

Accordingly, the sea, the locus of chaos, will ultimately cease to exist.[54] Until "that day" we are to live within an untamed world. For now, the boundary between chaos and order will be unpredictable and subject to times of stability as well as moments of violent disorder. This leads to a healthy respect and fear of the power of chaos within the natural order (see Ps 46:2–3; 107:23–30).

To summarize, in the Wisdom Tradition, chaos is portrayed as a turbulent sea or personified as a monster that no one other than God can tame. This is very different from uniform disorder or static randomness. The texts support the view that God has sovereignly chosen not to eliminate chaos (yet), as, presumably, this would not lead to the kind of cosmos that God intended. Why? Because order requires chaos, you cannot have one without the other. Indeed, perfect order would be boring and would not give rise to creativity, spontaneity, or development. Chaos and chance can also bring

---

52 See also 2 Esd 6:52: "But to Leviathan you gave the seventh part, the watery part; and you have kept them to be eaten by whom you wish, and when you wish."

53. I suggest that one should view the incidents where Jesus walks on water (Matt 14:22–33; Mark 6:45–52; John 6:16–21) and Jesus calming the storm (Mark 4:35–41; Luke 8:22–25; Matt 8:23–27) in this theological context. Taken as genuinely historical events, they were signs of the presence of the kingdom of God in a new and powerful way being manifest in Jesus Christ. It was not a publicity stunt to command allegiance; he already *had* the disciples' allegiance. Rather it was an unforgettable teachable moment that revealed to the disciples exactly *who* it was they were following.

54. See Anderson, *Contours of Old Testament Theology*, 301.

about good change, new possibilities, not just destruction. Yet, *unconfined* chaos is too tempestuous to allow, since the conditions necessary for the emergence of order and life are too fragile (cf. the great flood).[55] God gives freedom for chaos to be the "other" only within certain boundaries.

God's Creation in Job

Having reviewed the usage of the sea and its monsters in the Old Testament, let us briefly return to Job. The book of Job is both profound and enigmatic. Scholars have wrestled with its poetic contents and its bizarre prose prologue and surprising epilogue. There are diverse views on many aspects of the story, including the author's description of God's two speeches at the conclusion.[56] While this is not the place to explore the depths of this book, or the wider problem of suffering, *creation* is, as we have already seen, a persistent theme within Job.[57]

In chapters 38–41 God finally responds to Job in a somewhat incongruous way by presenting him with a tour of the natural world. Many readers (and scholars) would claim that God seems to be insensitive to Job and his suffering, and God's response seems to avoid the issues of Job's complaint.[58] Certainly the response is not what Job—or the reader—expected. Fretheim takes a more positive route, one that I find coherent and insightful. God's response is one that genuinely addresses Job's concerns, and is focused on nature because therein lies a key point that God wants Job to appreciate. After all, two of Job's original calamities were natural disasters (1:16, 19). God informs Job that he does not understand the way in which God's world works. Job interprets the disorder within nature as defective and/or mismanaged creation, rather than precisely the kind of world that God intended. Consequently, although the world is good, well-ordered, and reliable, it is also wild, untamed, and not risk-free to humankind. God, then, challenges Job to recognize the proper nature of the creation, and that

---

55. In the flood narrative, the rain ceases because God "restrains" the heavens (Gen 8:2). In the covenant with Noah, God does not eliminate chaos but simply promises that the regular cycles of nature will faithfully continue forever (Gen 8:22).

56. God speaks out of a whirlwind (38:1), i.e., theophany: see also Ezek 1:4; Nah 1:3; Zech 9:14. There is also a parallel with Isa 40:12–26, esp. 12–14.

57. Fretheim, *Relational Theology of Creation*, 219.

58. Murphy, *Tree of Life*, 42–43; Anderson, *Contours of Old Testament Theology*, 281–82; Birch et al., *Theological Introduction to the Old Testament*, 412.

suffering may be experienced in just such a world, quite apart from sin and evil. In so doing, Job may better appreciate what his place and role is within God's world, even in the midst of suffering.

God's first reply (38:1—40:2) is an exhaustive catalogue of his creative and sustaining acts. That speech can be divided into two sections: (a) cosmic and the physical order (38:4–38), and (b) God's providence for *wild* animals 38:39—39:30 (namely: the lion, raven, mountain goat, deer, wild donkey, wild ox, ostrich, war horse, hawk, and eagle). Like Genesis 1, it is not just the regions that God defines, but also what goes on within them. The writing style is a series of rhetorical questions, typical of wisdom literature, to which the implied answer is "no." This serves to highlight human ignorance and powerlessness in contrast to the extensive and complex creation that God created and continually sustains. These questions put Job in his place as someone who has "words without knowledge" (38:2) and yet who dares to argue with God (40:2). God's second speech (40:6—41:34) has a strong emphasis on two mysterious creatures, Behemoth and Leviathan. While these two animals could be thought to refer to the hippopotamus and crocodile, respectively, their darker, symbolic reputation cannot be overlooked—particularly in the context in which Leviathan and the sea has been used earlier within Job. Nevertheless, by taking these mythical beings as representatives of "chaos" does not make them—or the disorder they depict—morally evil. They are simply a part of the diverse and wonderful world that God has created. Still, chaos is truly awesome and beyond any human control. Fretheim asserts that it is not helpful to suggest that chaos is fully within divine control. While God has set a boundary to Leviathan's activity, that limit does not entail divine micromanagement. Rather, God lets his creatures function freely within their divine restrictions. What this reveals is that there are elements of God's good creation that are complex and ambiguous—not everything is neat and tidy, as Job presupposes it should be.[59] Fretheim concludes:

> This creational being and becoming is well-ordered, but the world does not run like a *machine*, with a tight causal weave; it has elements of *randomness* and chaos, of strangeness and wildness. Amid the order there is room for *chance*. . . . Given the communal character of the cosmos—its basic interrelatedness—every creature will be touched by the movement of every other. While this has negative potential, it also has a positive side, for only then is

---

59. Fretheim, *Relational Theology of Creation*, 235, 237.

there the genuine possibility for growth, creativity, novelty, surprise, and serendipity.[60]

In summary, a morally neutral chaos has a creative place within God's world, with both the potential for good and bad for humankind. Our dynamic world is not risk-free for humans—even for righteous people, such as Job.[61] God has made a world with a significant element of chaos and disharmony that are an integral and essential part of a world that is in the process of "becoming." Volcanoes are needed to replenish our atmosphere in order to sustain life; this requires a planet with an active geology. The earth has plate tectonics with earthquakes and tsunamis. Our sun-heated atmosphere sustains life, but it also gives rise to hurricanes, tornadoes, and cyclones. These messy, disorderly natural disasters have a role to play in our dynamic world. Order and chaos are inseparable; physical violence and the birth-death-decay cycle are features of God's good world. Yet these events also have the capacity to bring suffering to humans and animals. While untamed chaos has a God-ordained place within creation, God nevertheless declares this as "good."

CREATIO EX NIHILO AND CREATIO EX MATERIA

The traditional Christian doctrine of *creatio ex nihilo* is that God *freely* first created matter and then formed the cosmos from it.[62] This being the case, matter itself has all the inherent properties God intends. Moreover, God was not constrained or opposed (in any way) in the choice of the basic building blocks of the universe, or in how God ordered them—or, indeed,

---

60. Ibid., 244, emphasis mine. He adds: "And so God takes responsibility for Job's suffering by having created, and still sustaining, a world that is not risk-free and in which people can suffer undeservedly. The divine relationship to this world is such that God no longer acts with complete freedom, but from within a committed relationship to the structures of creation to which God will be faithful" (ibid., 244–45). Furthermore, he states: "And so God will sustain such an ordered and open-ended creation even in the face of the suffering ones who wish that God would have created a world wherein human beings could be free from suffering. That is a price, sometimes a horrendous price, which creatures pay for the sake of having such a world; but it is also the price that God pays, for God will not remove the divine self from that suffering and will enter deeply into it for the sake of the future of just such a world" (ibid., 237).

61. Fretheim, *Creation Untamed*, 81–84, 108.

62. A defense of *creatio ex nihilo* and a detailed study of Gen 1:1–2 is given in Copan and Craig, *Creation out of Nothing*, 29–70. See also see Fergusson, *Creation*, 15–35.

whether to create at all. This dual role of creation and formation is traditionally taken as demonstrating the absolute omnipotence of God over creation.[63]

There is, as we have just seen, a strong biblical case for God's continual restraining of chaos and keeping it within its bounds. Process theologians take this a step further and also infer that God created the cosmos out of preexisting material: *creatio ex materia*, or creation-out-of-chaos.[64] Indeed, the NRSV renders Genesis 1:1 not as an *absolute* beginning and it can be argued that creation out of preexisting material is also implied:

> In the beginning *when* God created the heavens and the earth, the earth *was* a formless void and darkness *covered the face of the deep*, while a wind from God swept over the face of the waters.[65]

The same sentiment is expressed by the writer of Wisdom 11:17: "For your all-powerful hand, which created the world out of formless matter...." The biblical case for creation out of preexistent material does not rest solely on these few verses in Genesis 1, whose words are—to most scholars—somewhat ambiguous. Combining pertinent texts from wisdom literature and the prophets with those of Genesis 1–3 presents a broader view of God's creative activity, as we have seen. Given all the biblical evidence, it is entirely reasonable to conclude that *creatio ex nihilo* was not high on the agenda of the Old Testament authors and redactors. There is, however, good, late evidence in 2 Maccabees 7:28 that is often seen as the first clear statement supporting creation-out-of-nothing (emphasis added):

> I beg you, my child, to look at the heaven and the earth and see everything that is in them, and recognize that *God did not make them out of things that existed*. And in the same way the human race came into being.

It is often pointed out that John 1:3, Romans 4:17, Colossians 1:16, and Hebrews 11:3 can be interpreted in the same way. Most scholars regard *creatio ex nihilo* as being first formulated in Christian thought in the late

---

63. E.g., see Migliore, *Faith Seeking Understanding*, 100, 407. For an excellent summary and discussion of the doctrine of creation, see 92–116.

64. See, e.g., Griffin, "Creation, Chaos and Evil," 108; Fretheim, "Genesis," 356; Fretheim, *Creation Untamed*, 20–24.

65. Emphasis mine; Gen 1:1 clearly indicates a *temporal* beginning, even if it is *not* an *absolute* beginning.

second century CE.[66] Although, it is fair to say, deeming *creatio ex nihilo* as a "post-biblical" addition is not a particularly strong argument in itself since the doctrine of the Trinity was not formalized until the Nicene-Constantinopolitan Creed at the 381 CE ecumenical council.

One of the reasons that even questioning the traditional doctrine of *creatio ex nihilo* is contentious to some Christians is its close connection with God's sovereignty and omnipotence. There is, however, an important category difference between the two: sovereignty is a status and omnipotence is a capability. Only God's ability to rule the cosmos is threatened by chaos, not God's right to kingship. In contrast to the Genesis 1–2 accounts, which emphasize God as omnipotent *Creator*, the wisdom texts bring a healthy counterbalance by giving suitable stress to God as *Sustainer*. God's covenant with Noah also presents God in this way (see Gen 8:20—9:17). The term "Sustainer" should not be understood as *preserving* the original creation, as if it were to be maintained in a static or equilibrium state. Rather, God is sovereign over the *continuing* creation, enabling it to become all it can become, to fulfill its original potential, and—ultimately—to bring creation to completion at the eschaton.[67] Within all the above biblical texts, there is no doubt concerning God's capability to achieve God's final goal, despite the present persistence of chaos, so God's omnipotence and sovereignty are ultimately not compromised.

As discussed earlier, Leviathan is a creature that was declared "very good" (Gen 1:21, 31). Demythologized chaos is nevertheless *real*; a "force" to be reckoned with. But chaos, *per se*, should not be seen as inherently evil in a moral sense, nor an enemy of God, rather simply having the capability to oppose order.[68] *Creatio ex materia* affirms that chaos is morally neutral. One can agree with that assertion and at the same time support *creatio ex nihilo*, since nothing God has made is inherently evil.

---

66. Griffin, "Creation, Chaos and Evil," 109–14; McGrath *Christian Theology*, 219–20; Fergusson, *Creation*, 15. Griffin raises a legitimate concern: If God is said to have created the world out of absolute nothingness then the origin of the evil cannot be explained, at least not without implying that God's goodness is less than perfect. Creation out of pre-existing material on the other hand, gives rise to a possible explanation for the origin of evil while defending God's total goodness. Griffin, "Creation, Chaos and Evil," 112, 114.

67. Fretheim, *Relational Theology of Creation*, 3–9.

68. Fretheim, *Creation Untamed*, 12–25.

SUMMARY AND CONCLUSION

This chapter has reviewed some of the key Old Testament creation texts. There are, of course, others too (e.g., Prov 8:22–31; and frequent references in Isa 40–55), but they can also—broadly speaking—fit into Levenson's twin categories. Little can compare to the resplendent majesty of Genesis 1, but it is important to consider the whole canon of Scripture in forming a theology of creation. The contrast between the two creation stories in Genesis is self-evident. Add to that the vivid imagery in the Wisdom Tradition and we see a rich tapestry of storytelling and poetic metaphors that all glorify the Creator—as illustrated in Psalm 148 at this chapter's opening. The Fall narrative of Genesis 3 has also been explored. Since it is not literal history, we need not tie ourselves in knots over its portrayal of physical death. Nevertheless, the *theological* importance of the story is not diminished since it is a foundational narrative for redemptive history. The often overlooked theme of the sea and chaos has also been reviewed giving added—and needed—emphasis on God's sustaining activity and continual creation.

Our issues surrounding origins are inevitably connected with questions pertaining to ultimate destiny. Within the Christian message is the powerful hope for the future cosmic redemption that has already been realized in Jesus Christ. The resurrection of Jesus is the pivotal historical event that creates universal hope for the future.[69] Of course much of that future is shrouded in mystery both theologically and scientifically.[70] It is safe to say that the new creation will *not* be a return to the original beginning. If that were the case, everything that has happened in between would be of no consequence.[71] What we do know—from the resurrection—is that the new creation will be in some way continuous with this present creation; it will not be made "out of nothing" or from scratch. This seems to be the way of God whose new covenants are both continuous and discontinuous with previous ones. And God is a God who keeps his promises. So we rely on the faithfulness of God towards humankind and all of his creation and in his ability to accomplish all that he desires to achieve in history.

---

69. This is explored in Wright, *Surprised by Hope*.

70. This is explored in Polkinghorne, *God of Hope*, and Wilkerson, *Christian Eschatology*.

71. Fretheim, *Relational Theology of Creation*, 9.

Questions were posed at the beginning of this chapter: Is it *possible* to harmonize early Genesis with the findings of science? Should we even try? My response is evident in all the above discussion: *"Let Scripture be."* Those who proclaim a high emphasis for *sola scriptura* should welcome such a stance. Yet we need to appreciate the varied genres of the biblical literature along with recognition that it is a collection of books written at different times and places, and for different audiences and contexts. Canonical criticism, mentioned in chapter 2, values the entire Bible and recognizes the interconnectedness in the final form of its contents. Even so, it is a critical view of the texts—one that respects the integrity of the parts as well as the whole. In the discussion above I have endeavoured to follow that spirit and to integrate what emerges from the exegesis. Of course I am not unbiased; I bring my own theological lens and personal history and background into this process. Part of my stated plea is *not* to attempt to read modern science into Scripture, but to recognize the prescientific nature of these sacred writings. In this sense I am advocating for, in Barbour's terms, an "independence" stance between science and *Scripture* while endorsing a "dialogue" perspective between science and *theology*. This allows us to embrace biblical stories on their own merits, together with Scripture's overarching narrative or message. This, in turn, requires us to hold to the canonical whole of Scripture. Healthy dialogue can lead to the integration of science and theology, and to new understandings of traditional Christian doctrines so enhancing our faith in the One who gives life.

Afterword

> They say a little knowledge is a dangerous thing, but it's not one half so bad as a lot of ignorance. —Terry Pratchett

As this book comes to a close, I am reminded of a well-known line from one of Churchill's wartime speeches: "This is not the end. It is not even the beginning of the end. But it is, perhaps, the end of the beginning." This sentiment is most appropriate as this study has, primarily, explored foundations and frameworks for moving forward in faith. The conversation between science and Christianity continues, of course, both in terms of going deeper into specific matters and in addressing completely new topics. In addition to physics, my own area of expertise, all the other scientific disciplines have important contributions to make in this ongoing dialogue. Together this allows for timely issues in science and medicine to be addressed, like: climate change, genetic engineering, sustainable energy, assisted dying, abortion, animal welfare, environmental pollution, mining techniques, space exploration, etc. Much has already been discussed and debated in these areas (see bibliography); this book is simply an appetizer, one that *orients* our thinking. My desire is that the result of these discussions not be merely one-way traffic, whereby the findings of science provide helpful insights to assist in formulating—or revising—theologies of nature. Rather, I hope that the directions in which certain scientific fields go will be enriched and shaped by theology and ethics. That latter vision is challenging, given the general discontinuity between academia and the church. Nevertheless, times are changing and a new social conscience is being awakened concerning humankind's abuse of nature simply out of greed and self-interest. In addition, ongoing advances in medicine remind us again and again of the importance of the ancient question: "What does it mean to be human?"

Afterword

The need for serious dialogue is *now*. One anticipated outcome would be to influence both the *priorities* and *methodologies* of scientific research and its applications.

This introductory work has covered a lot of ground. A canvas has been painted with a broad brush. We live in an age of "experts" where the fine details of a small aspect of the overall painting can be one's life work. Important though that may be, we also need to step back and not lose sight of the overall picture. This is much harder to do, of course. Not least as many in our postmodern world no longer believe there is such a picture—or at least not a unique one that is true for all. I am acutely aware of the danger in delving into fields outside one's area of expertise. Having been a researcher focusing on the details, one has to be "retrained" to recognize the existence of the bigger picture—i.e., *context*. As a Christian, I *do* believe that we are part of a canvas that God is painting. The picture is not a simple one because we also contribute to the outcome. The canvas contains multiple layers. And each of us, and the traditions of which we are a part, tend to focus on the layer of the picture most evident to our time and culture. Nevertheless, God—the master creative Artist—has an overall image in mind, while being flexible with the details.

This study has been undertaken in a spirit of humility and with full self-awareness that I am no polymath. Nevertheless, it arises out of study and reflection spanning over thirty-five years. As is the way of such things, I wish I knew at the outset what I know now—but I am grateful for my continuing faith journey that shapes who I am. My hope is that this work will be helpful to you in your own adventure of faith.

Appendix 1

Traditional Theistic Arguments for the Existence of God

Natural theology has long sought for reasoned arguments for the existence of a deity starting from logic and nature. This quest continues to be of value in various Christian traditions as a complement to a *prior* commitment to faith, rather than an agnostic's search of rational grounds *for* faith. If anything, as we shall see, the search establishes a cold "god of the philosophers" rather than the personal God of the Christian tradition who is supremely revealed in Jesus the Christ. Natural theology is closely associated with the classic arguments of first cause (cosmological) and design (teleological), discussed extensively by Aquinas.[1] These arguments are briefly outlined and critiqued in this appendix.

---

1. There is also the ontological argument, first developed by St. Anselm (1033–1109), which begins with the emphasis on the *idea* of God as "a being which nothing greater (more perfect) can be *conceived*." Anselm then pointed out that this being must exist in reality and not simply be conceived in the human mind. His reasoning was that if God was only confined to the mind then a greater being would be one who *actually* existed in reality—hence God exists! However, just because I can imagine something that is wonderful does not automatically mean it *must* exist, as existence is not a straight forward property of an object. The argument can be re-phrased to say that God must uniquely *necessarily* exist, rather than be contingent. But this is only valid if it is true, i.e., that God's existence is necessary, which is precisely what an atheist doubts. Only *if* there is a God, must he exist. Notice that this argument is based on logic, rather than science or theology, and therefore is a philosophical argument. Logical proofs on such matters are generally illusive, but potentially useful if considering the *possibility* that God exists. Recall, however, that Anselm also said: "For I do not seek to understand that I may

Appendix 1

We recognize that the universe itself is contingent, in other words the cosmos does not necessarily have to exist and it could exist differently. This connects with Hawking's question: "What is it that breathes fire into the equations and makes a universe for them to describe?" or "Why is there something rather than nothing?"[2] This is the starting point for the cosmological argument, which has two main variants, temporal and non-temporal (or logical). The more familiar argument is temporal. Everything that begins to exist has a *cause* for its existence; the universe began to exist, therefore the universe has a cause for its existence. This argument can be adapted into the non-temporal language of contingency and necessity. That version ultimately asserts that the universe, being contingent, requires a necessary (i.e., non-contingent) agency to cause its existence. Theists call this necessary being "God."

Despite the big bang being experimental evidence for the beginning of our universe, some critics might object by saying that the cosmos is eternal and therefore the number of causes is infinite. One way for a universe with a finite origin to appear infinite is to argue for no special, unique beginning, but a universe oscillating between a big bang and big crunch. Regardless of the possible veracity of this suggestion, critics have responded by saying that you cannot have an infinite number of causes in *reality*, only within mathematics, else physical absurdities will arise. In addition, even if an infinite set of causes were actually possible, it cannot explain something's existence. If each causal condition is contingent, so would it be for an infinite series.

Some critics might further object on the grounds that quantum mechanics contains the notion of no intelligible or discernible cause for quantum events. Quantum indeterminacy is certainly a challenge to the cosmological argument, due to its emphasis on causation.[3] However, the lack of an intelligible cause is not a free lunch; a quantum particle does not arise out of literally nothing. Even a vacuum to a physicist is a hive of energetic activity and potentiality. Moreover, it is not obvious that you can apply this argument to the universe as a whole.

---

believe, but I believe in order that I may understand." All knowledge, including theology, begins with faith or trust, rather than perpetual skeptical uncertainty and doubt, and uses human reason in the service of furthering an understanding of the grounds for faith, and its object. For further reading, see, e.g., Peterson et al., *Reason and Religious Belief*, 92–96, and Evans and Manis, *Philosophy of Religion*, 63–67.

2. Hawking, *Brief History of Time*, 174.
3. Peterson et al., *Reason and Religious Belief*, 97.

Atheist philosopher Bertrand Russell said that even if all the parts of the universe are contingent you cannot infer that the universe itself is contingent.[4] He concluded that we cannot ask about the cause of the universe, "it's just there and that's all."[5] This view can be challenged; if *all* the parts of the universe—both matter and energy—ceased to exist simultaneously, then the universe itself would cease to exist. *If* the universe can cease to exist then it must be contingent, and therefore requires an explanation for its very existence.

What are we to conclude from this brief foray into logic? Is the universe just a brute fact? Can it be that no reason can be given to Leibniz's profound question: "Why there is something rather than nothing?" Is the universe therefore ultimately unintelligible?[6] Regardless, the cosmological argument is evidently not convincing to all. But the rejection of the argument implicitly carries with it a commitment to a rival philosophical perspective.[7] The question then is: "Which philosophical system is most plausible?" While the cosmological argument cannot be used to *prove* the existence of "God," it does demonstrate that belief in God is not irrational. Moreover, if you abandon the quest for logical proof, you could perhaps go further and claim that "God" is the *most probable* cause for the existence of the universe.

The teleological ("directed toward a goal" or "purposeful") argument is more commonly known as the "argument from design." This argument, like the previous one of first cause, is also ancient, being found in Aristotle's final cause and, although reformulated by Aquinas, it is more often associated with William Paley (1743–1805). Paley likened the apparent order within the universe to that of a newly discovered watch. Having found the item you would, after perceiving the cogs and regular internal workings, assert that it was *designed* and not attribute it to mindless or purposeless chance. This would be true even though you had no idea of its purpose, or whether it successfully met that requirement, or even if after closer scrutiny you could not understand its mechanism. It is worth noting that Paley's context was that of the Industrial Revolution and so, for him, mechanism

---

4. This is referred to as the "fallacy of composition" in philosophy. However, inferences of the whole from the parts are not always fallacious; see Evans and Manis, *Philosophy of Religion*, 71.

5. Peterson et al., *Reason and Religious Belief*, 104.

6. Hick, *Philosophy of Religion*, 22.

7. Evans and Manis, *Philosophy of Religion*, 76.

implied contrivance—intentionality, not simply random chance.[8] For Paley, biological systems, like the human heart or eye, were similarly "contrived" and were therefore designed with a purpose in mind by an intelligent Designer. This idea is then extended to the universe as a whole, an inference which is widely viewed as problematic.[9] Even so, critics have pointed out that this is not really a fair analogy, as the universe is more like a developing organism rather than a mechanistic device. Nature appears to self-organize and adapt itself, given enough time, with natural selection acting as the organizing principle. Evolutionary naturalism seems to have provided a reasonable alternative for describing the end-means (*telos*) within nature, without the necessity of an external Designer.[10] Be that as it may, evolution, *per se*, certainly does not preclude the possibility of theistic design, as some might want to assert. On the contrary, evolution—either cosmic or biological—could be the process by which the Designer ("God") realizes his purposes within the universe. Nevertheless, natural selection does significantly weaken the argument for the need of a Designer. At best the teleological argument makes the notion of an external Designer plausible.

---

8. McGrath, *Science and Religion*, 100–101.
9. See footnote 4 in this appendix.
10. Peterson et al., *Reason and Religious Belief*, 106.

Appendix 2

A Brief Excursion into Metaphysics

The traditional scientific method and the contributions of Popper, Kuhn, and Polanyi to the nature of science were discussed in chapter 3. Another approach to that topic is to ask the question—one that is akin to Aristotle's final cause—"What is science *for*?" One response is to say that the ultimate goal of science is *prediction*, resulting in the power to manipulate matter. This is a utilitarian view of science in which the question "Does it work?" is the only criteria for success. If this were the case, the argument between Galileo and Bellarmine (see chapter 1) was pointless because they were addressing the question "Is this a true description of reality?" The first view is that of instrumentalism and the second is realism. Scientists, however, are interested in more than mere prediction, they also want *understanding*. Polkinghorne writes:

> I have never known anyone working in fundamental physics who was not motivated by the desire to comprehend better *the way the world is*. It is because they yield *understanding*, though often having low or zero predictive power, that theories of origins, such as cosmology or evolution, are rightly classed as parts of science.[1]

That goal of understanding has, however, to be qualified. Where does understanding come from? Is it an invention of humankind or is it determined by the nature of the world with which we interact?

---

1. Polkinghorne, *One World*, 20–21, emphasis mine.

Appendix 2

Idealists take the view that "reality" is primarily a property of the mind. Indeed, the extreme position asserts that there is no real external world; it only exists in the conscious mind. From this perspective, science should be a branch of psychology! This was cleverly incorporated into the trilogy of *Matrix* movies. The apparent order of the "world" we think we perceive is simply as result of the way we observe it. This implies X-rays, electrons, and nucleons did not even exist until they were "discovered." The question then is: "Are these discoveries particular to the individual who found them (nominalism), or do they have a universal quality about them"? There is no self-evident answer, but nominalism would be a disastrous basis for any corporate study of nature. And idealism isn't much better; it is too weak a basis to give confidence or cause to *instigate* science, let alone motivating people to maintain and develop the enterprise. Euan Squires concludes: "I believe that [Idealism] is logically unassailable but, in practice, foolish and sterile."[2]

If we accept that there is a reality external to the conscious mind, then we are left with the two broad alternatives of instrumentalism and realism. These metaphysical questions understandably arise when wrestling with the adoption of a new paradigm, such as that of modern physics. As we have seen, the instrumentalist regards science purely as a means to an end. Good theories are successful tools for predicting and numerically modelling phenomena, and thereby used to control nature, but they are not regarded as "explanations" of reality. Consequently, it is meaningless to speak of a "true" or "false" theory, as science makes no pretense to address the actual nature of reality, only to *model* it. The conceptual challenge of the quantum world, such as the uncertainty principle, made some scientists at the time wonder if the theory was simply a calculation procedure to model the phenomena. At one point, Bohr said privately to a friend:

> There is no quantum world. There is only abstract quantum physical description. It is wrong to think that the task of physics is to find out how nature *is*. Physics concerns what we can say about nature.[3]

One can have some sympathy with Bohr's comments, because, as Richard Feynman once said, "I think I can safely say that nobody really understands

---

2. Squires, *Conscious Mind*, 74.
3. Cited in Polkinghorne, *One World*, 44.

quantum mechanics"![4] Nevertheless, Chalmers criticizes this view as too cautious and defensive, and makes the fair point:

> The fact that theories can lead to novel predictions is an embarrassment for the instrumentalists. It must seem a strange kind of accident to them that theories, that are supposed to be mere calculating devices, can lead to the discovery of new kinds of observable phenomena by way of concepts that are theoretical fictions.[5]

Such new discoveries have certainly arisen in the case of quantum mechanics.

In contrast to instrumentalism, "critical realists view theories as partial representations of limited aspects of the world as it interacts with us."[6] Abstract models, though tentative, are genuine attempts to imagine the structures of the world that give rise to the interactions observed in experiments. If we regard science as a "quest for truth" (which the realist's position implies) then a further useful notion, introduced by Popper, is the idea of an "approximation to the truth." Newton's mechanics, which replaced earlier theories of motion, has been superseded by quantum mechanics and relativity. But despite such falsifications Popper would say that we have progressed ever closer to "the truth." Obviously we do not know for sure if our current theories are true in any absolute sense, but we believe them to be better than their predecessors, and so we can speak of a slow convergence toward the truth. Popper termed this "verisimilitude." Clearly our understanding of nature will never become complete, as there will always be new phenomena to explore. In addition to that qualification, discontinuities can arise—like those associated with a Kuhnian paradigm shift—which interrupts the underlying convergence process. Critics, understandably, pounce on that point. Nevertheless, *critical* realism acknowledges the findings of science are subject to revision, as well as the role of the observer, and the use of personal judgement in science.[7]

It is not only scientists who generally adopt a realistic interpretation to their endeavors, so do the general public. We teach science in our schools *as if* it were true: it is assumed that science tells us how the world actually is. It is for this reason there is conflict between biblicism and scientific

---

4. Feynman, *Character of Physical Law*, 129.
5. Chalmers, *What Is This Thing Called Science?* (2nd ed.), 149.
6. Barbour, *Religion and Science*, 168.
7. Polkinghorne, *One World*, 22–23.

materialism, which also fuels controversy over teaching creationism and evolution in schools in some parts of the United States. The debate would not be contentious if everyone adopted instrumentalism—not that I am advocating that stance.

As an experimental physicist, the critical realist's position makes eminent common sense, even if it is—like all positions—unprovable to the philosopher of science.[8] To the critical realist, then, there is an assumption that epistemology is closely related to ontology—in other words, that what we *know* about the world, albeit provisionally, correlates with the way the world actually *is*. This viewpoint is particularly relevant when discussing the implications of Heisenberg's uncertainty principle.

---

8. E.g., see the discussion in Chalmers, *What Is This Thing Called Science?*, 226–46.

Bibliography

Anderson, Bernhard W. *Contours of Old Testament Theology*. Minneapolis: Fortress, 2011.
Anderson, G. A. "Eden, Garden of." In *New Interpreter's Dictionary of the Bible*, 2:186-87. Nashville: Abingdon, 2007.
Appleyard, Bryan. *Understanding the Present: Science and the Soul of Modern Man*. London: Picador, 1992.
Aquinas, Thomas. *Summa Contra Gentiles*. Translated by Joseph Rickaby. www.catholicprimer.org/aquinas/aquinas_summa_contra_gentiles.pdf.
———. *Summa Theologica*. Benziger ed., 1947. Translated by Fathers of the English Dominican Province. http://www.ccel.org/ccel/aquinas/summa.
Augustine. *On Christian Doctrine*. http://www.ccel.org/ccel/augustine/doctrine.xix_1.html.
Baelz, Peter R. *Does God Answer?* London: Darton, Longman & Todd, 1982.
Balserak, Jon. "Exegesis and *Doctrina*." In *The Calvin Handbook*, edited by H. J. Selderhuis, 372-78. Grand Rapids: Eerdmans, 2009.
Barbour, Ian G. *Religion and Science: Historical and Contemporary Issues*. New York: Harper Collins, 1997.
———. *When Science Meets Religion*. London: SPCK, 2000.
Barrow, John D., and Frank J. Tipler. *The Anthropic Cosmological Principle*. Oxford: Oxford University Press, 1986.
Bartholomew, David J. *God, Chance and Purpose: Can God Have It Both Ways?* Cambridge: Cambridge University Press, 2008.
Barton, Stephen C., and David Wilkinson, eds. *Reading Genesis After Darwin*. Oxford: Oxford University Press, 2009.
Basinger, David, and Randall Basinger, eds. *Predestination and Free Will: Four Views on Divine Sovereignty and Human Freedom*. Downers Grove: InterVarsity, 1986.
Beilby, James K., and Paul R. Eddy, eds. *Divine Foreknowledge: Four Views*. Downers Grove: InterVarsity, 2001.
Bellarmine, Roberto. *Disputations*. Vol. 1, Controversy 1, *On the Word of God, Written and Unwritten*. http://www.aristotelophile.com/current.htm.
Berry, R. J. *God's Book of Works: The Nature and Theology of Nature*. London: T. & T. Clark, 2003.
Berry, R. J., and T. A. Noble, eds. *Darwin, Creation, and the Fall: Theological Challenges*. Nottingham: Apollos, 2009.
Birch, Bruce C., et al. *A Theological Introduction to the Old Testament*. 2nd ed. Nashville: Abingdon, 2005.

BIBLIOGRAPHY

Bird, Phyllis A. "The Authority of the Bible." In vol. 1 of *New Interpreter's Bible Commentary*, edited by Leander E. Keck, 33-64. Nashville: Abingdon, 1994.

Blackwell, Richard J. *Galileo, Bellarmine, and the Bible*. Notre Dame: University of Notre Dame Press, 1991.

Blocher, Henri. *In the Beginning: The Opening Chapters of Genesis*. Translated by David G. Preston. Downers Grove: InterVarsity, 1984.

Bonhoeffer, D. *Creation and Fall: A Theological Exposition of Genesis 1-3*. Minneapolis: Fortress, 1997.

Boring, M. Eugene. "Matthew." In vol. 8 of *New Interpreter's Bible Commentary*, edited by Leander E. Keck. Nashville: Abingdon, 1994.

Boyd, Gregory. A. *God at War: The Bible and Spiritual Conflict*. Downers Grove: InterVarsity, 1997.

———. *God of the Possible: A Biblical Introduction to Open Theism*. Grand Rapids: Baker, 2000.

Brooke, John Hedley. *Science and Religion: Some Historical Perspectives*. Cambridge: Cambridge University Press, 1991.

Brown, Colin. *Miracles and the Critical Mind*. Pasadena: Fuller Seminary Press, 2006.

Bruce, F. F. *The Canon of Scripture*. Downers Grove: InterVarsity, 1988.

Brueggemann, Walter. *Genesis*. Interpretation: A Bible Commentary for Teaching and Preaching. Louisville: John Knox, 1982.

Calvin, John. *Commentary on the Psalms*. Vol. 1. Translated by J. A. Edinburgh. http://www.ccel.org/ccel/calvin/calcom08.html.

———. *Commentary on the Psalms*. Vol. 5. Translated by J. A. Edinburgh. http://www.ccel.org/ccel/calvin/calcom12.html.

———. *Institutes of the Christian Religion*. Translated by Henry Beveridge. http://www.ccel.org/ccel/calvin/institutes.html.

Carlson, Richard F., ed. *Science and Christianity: Four Views*. Downers Grove: InterVarsity, 2000.

Carlson, Richard F., and Tremper Longman III. *Science, Creation, and the Bible: Reconciling Rival Theories of Origin*. Downers Grove: InterVarsity, 2010.

Chalmers, Alan F. *What Is This Thing Called Science?* 2nd ed. Buckingham: Open University Press, 1982; 3rd ed., 1999.

Charlesworth, J. H. "Paradise." In *New Interpreter's Dictionary of the Bible*, 4:377-78. Nashville: Abingdon, 2009.

Chicago Statement on Biblical Inerrancy. http://www.bible-researcher.com/chicago1.html.

Childs, Brevard S. *Biblical Theology of the Old and New Testaments: Theological Reflection on the Christian Bible*. Minneapolis: Fortress, 2011.

Cobb, John B., Jr., and Clark H. Pinnock, eds. *Searching for an Adequate God: A Dialogue between Process and Free Will Theists*. Grand Rapids: Eerdmans, 2000.

Collins, Francis S. *The Language of God*. New York: Free Press, 2007.

Conrad, E. W. "Satan." In *New Interpreter's Dictionary of the Bible*, 5:112-16. Nashville: Abingdon, 2009.

Cootsona, Greg. "When Science Comes to Church." *Christianity Today*, March 5, 2014. http://www.christianitytoday.com/ct/2014/march-web-only/when-science-comes-to-church.html.

Copan, Paul, and William Lane Craig. *Creation out of Nothing: A Biblical, Philosophical, and Scientific Exploration*. Grand Rapids: Baker, 2004.

Bibliography

Cotter, Wendy. "Miracle." In *New Interpreter's Dictionary of the Bible*, 4:99–106. Nashville: Abingdon, 2009.

Craig, William Lane. "The Middle Knowledge View." In *Divine Foreknowledge: Four Views*, edited by James K. Beilby and Paul R. Eddy, 119–43. Downers Grove: InterVarsity, 2001.

———. *Reasonable Faith: Christian Truth and Apologetics*. 3rd ed. Wheaton, IL: Crossway, 2008.

Davies, Brian. *An Introduction to the Philosophy of Religion*. 3rd ed. Oxford: Oxford University Press, 2004.

Davies, Paul. *The Mind of God: Science and the Search for Ultimate Meaning*. London: Penguin, 1992.

Davis, Stephen T., ed. *Encountering Evil: Live Options in Theodicy*. 2nd ed. Louisville: Westminster John Knox, 2001.

Day, J. N. "God and Leviathan in Isaiah 27:1." *Bibliotheca Sacra* 155 (1998) 423–36.

Dembski, William A. *The Design Revolution: Answering the Toughest Questions about Intelligent Design*. Downers Grove: InterVarsity, 2004.

———. *Intelligent Design: The Bridge between Science and Theology*. Downers Grove: InterVarsity, 1999.

Dillenberger, J. *Protestant Thought and Natural Science*. Westport: Greenwood, 1977.

Donovan, Vincent. J. *Christianity Rediscovered*. 25th anniv. ed. Maryknoll: Orbis, 2003.

Dunn, James D. G. "2 Timothy." In vol. 11 of *New Interpreter's Bible Commentary*, edited by Leander E. Keck. Nashville: Abingdon, 2000.

Enns, Peter. *The Bible Tells Me So: Why Defending Scripture Has Made Us Unable to Read It*. New York: HarperOne, 2014.

———. *The Evolution of Adam: What the Bible Does and Doesn't Say about Human Origins*. Grand Rapids: Brazos, 2012.

———. *Inspiration and Incarnation: Evangelicals and the Problem of the Old Testament*. Grand Rapids: Baker, 2005.

Evans, C. Stephen, and R. Zachary Manis. *Philosophy of Religion: Thinking about Faith*. 2nd ed. Downers Grove: InterVarsity, 2009.

Evans, Craig A., and Emanuel Tov, eds. *Exploring the Origins of the Bible: Canon Formation in Historical, Literary and Theological Perspective*. Grand Rapids: Baker, 2008.

Ferguson, Kitty. *The Fire in the Equations: Science, Religion, and the Search for God*. Philadelphia: Templeton, 1994.

Fergusson, David. *Creation*. Grand Rapids: Eerdmans, 2014.

Ferngren, G. B., ed. *Science and Religion: A Historical Introduction*. Baltimore: John Hopkins University Press, 2002.

Feynman, Richard P. *The Character of Physical Law*. London: Penguin, 1992.

Fretheim, Terence E. *Creation Untamed: The Bible, God, and Natural Disasters*. Grand Rapids: Baker, 2010.

———. "Genesis." In vol. 1 of *New Interpreter's Bible Commentary*, edited by Leander E. Keck. Nashville: Abingdon, 1994.

———. *God and World in the Old Testament: A Relational Theology of Creation*. Nashville: Abingdon, 2005.

———. "Is Genesis 3 a Fall Story?" *Word & World* 14 (1994) 144–53.

Galilei, Galileo. *Letter to the Grand Duchess Christina*. In *Discoveries and Opinions of Galileo*, edited by Stillman Drake, 173–216. New York: Anchor-Doubleday, 1957. http://inters.org/galilei-madame-christina-Lorraine.

BIBLIOGRAPHY

Ganssle, Gregory E., ed. *God and Time: Four Views*. Carlisle: Paternoster, 2002.
Gerrish, B. A. "The Reformation and the Rise of Modern Science." In *The Impact of the Church upon Its Culture: Reappraisals of the History of Christianity; Essays in Divinity*, edited by J. C. Brauer, 2:231–65. Chicago: Chicago University Press, 1968.
Gingerich, Owen. *The Book Nobody Read: Chasing the Revolutions of Nicolaus Copernicus*. New York: Walker, 2004.
Godfrey-Smith, Peter. *Theory and Reality: An Introduction to the Philosophy of Science*. Chicago: Chicago University Press, 2003.
González, Justo L. "How the Bible Has Been Interpreted in Christian Tradition." In vol. 1 of *New Interpreter's Bible Commentary*, edited by Leander E. Keck, 83–106. Nashville: Abingdon, 1994.
Grant, Edward. *God and Reason in the Middle Ages*. Cambridge: Cambridge University Press, 2004.
———. *A History of Natural Philosophy: From the Ancient World to the Nineteenth Century*. Cambridge: Cambridge University Press, 2007.
———. *Science and Religion, 400 BC to AD 1550: From Aristotle to Copernicus*. Baltimore: Johns Hopkins University Press, 2004.
Green, Joel B. "Soul." In *New Interpreter's Dictionary of the Bible*, 5:358–59. Nashville: Abingdon, 2009.
Griffin, David Ray. "Creation out of Nothing, Creation out of Chaos, and the Problem of Evil." In *Encountering Evil: Live Options in Theodicy*, 2nd ed., edited by Stephen T. Davis, 108–25, 137–44. Louisville: Westminster John Knox, 2001.
Hall, Christopher A., and John Sanders. *Does God Have a Future? A Debate on Divine Providence*. Grand Rapids: Baker, 2003.
Hall, Douglas John. *God and Human Suffering: An Exercise in the Theology of the Cross*. Minneapolis: Augsburg, 1986.
Harrington, Daniel J. "Introduction to the Canon." In vol. 1 of *New Interpreter's Bible Commentary*, edited by Leander E. Keck, 7–21. Nashville: Abingdon, 1994.
Hasker, William. *The Triumph of God over Evil: Theodicy for a World of Suffering*. Downers Grove: InterVarsity, 2008.
Hawking, Stephen. *A Brief History of Time: From the Big Bang to Black Holes*. New York: Bantam, 1988.
Hick, John H. *Philosophy of Religion*. 4th ed. Cranbury: Pearson, 1989.
Hiebert, T. "Chaos." In *New Interpreter's Dictionary of the Bible*, 1:582–83. Nashville: Abingdon, 2006.
Hodge, B. C. *Revisiting the Days of Genesis*. Eugene, OR: Wipf & Stock, 2011.
Holladay, Carl R. "Contemporary Methods of Reading the Bible." In vol. 1 of *New Interpreter's Bible Commentary*, edited by Leander E. Keck, 125–49. Nashville: Abingdon, 1994.
Hooykaas, R. *Religion and the Rise of Science*. Edinburgh: Scottish Academic, 1972.
Hume, David. *Enquiry concerning Human Understanding*. http://www.earlymoderntexts.com/authors/hume.html.
Janzen, J. Gerald. *Job*. Interpretation: A Bible Commentary for Teaching and Preaching. Louisville: John Knox, 1985.
Jowers, Dennis W., ed. *Four Views on Divine Providence*. Grand Rapids: Zondervan, 2011.
Kinnaman, David. *You Lost Me: Why Young Christians Are Leaving Church . . . and Rethinking Faith*. With Aly Hawkins. Grand Rapids: Baker, 2011.

Bibliography

Kuhn, Thomas S. *The Copernican Revolution: Planetary Astronomy in the Development of Western Thought*. Cambridge: Harvard University Press, 1957.

———. *The Structure of Scientific Revolutions*. 3rd ed. Chicago: University of Chicago Press, 1996.

Langford, Jerome J. *Galileo, Science and the Church*. 3rd ed. Ann Arbor: University of Michigan Press, 1992.

Levenson, Jon D. *Creation and the Persistence of Evil: The Jewish Drama of Divine Omnipotence*. Princeton: Princeton University Press, 1988.

Lewis, C. S. "The Efficacy of Prayer." In *The World's Last Night and Other Essays*, 3–12. Orlando: Mariner, 2002.

———. *God in the Dock*. Edited by Walter Hooper. Grand Rapids: Eerdmans, 1970.

———. *Miracles: A Preliminary Study*. New York: HarperCollins, 2001.

Lindberg, Carter. *The European Reformations*. 2nd ed. Chichester: Wiley-Blackwell, 2010.

Lindberg, David C. *The Beginnings of Western Science: The European Scientific Tradition in the Philosophical, Religious, and Institutional Context, 600 B.C. to A.D. 1450*. Chicago: University of Chicago Press, 1992.

Lindberg, David C., and Ronald L. Numbers, eds. *God and Nature: Historical Essays on the Encounter between Christianity and Science*. Los Angeles: University of California Press, 1986.

Long, Thomas G. *What Shall We Say? Evil, Suffering, and the Crisis of Faith*. Grand Rapids: Eerdmans, 2011.

Mackay, Donald M. *The Clockwork Image*. London: InterVarsity, 1974.

———. *Science, Chance, and Providence*. Oxford: Oxford University Press, 1978.

Mays, James L. *Psalms*. Interpretation: A Bible Commentary for Teaching and Preaching. Louisville: John Knox, 1994.

McGrath, Alister E. *Christian Theology: An Introduction*. 5th ed. Chichester: Wiley-Blackwell, 2011.

———. *A Fine-Tuned Universe: The Quest for God in Science and Theology*. Louisville: Westminster John Knox, 2009.

———. *The Foundations of Dialogue in Science and Religion*. Oxford: Blackwell, 1998.

———. *Science and Religion: An Introduction*. Oxford: Blackwell, 1999.

McLean, B. H. *Biblical Interpretation and Philosophical Hermeneutics*. Cambridge: Cambridge University Press, 2012.

Migliore, Daniel L. *Faith Seeking Understanding: An Introduction to Christian Theology*. 2nd ed. Grand Rapids: Eerdmans, 2004.

Moltmann, Jürgen. *Science and Wisdom*. Translated by Margaret Kohl. London: SCM Press, 2003.

Newbigin, Lesslie. *Foolishness to the Greeks: The Gospel and Western Culture*. Grand Rapids: Eerdmans, 1986.

———. *The Gospel in a Pluralist Society*. Grand Rapids: Eerdmans, 1989.

———. *The Open Secret: An Introduction to the Theology of Mission*. Rev. ed. Grand Rapids: Eerdmans, 1995.

———. *Proper Confidence: Faith, Doubt, and Certainty in Christian Discipleship*. Grand Rapids: Eerdmans, 1995.

———. *Truth to Tell: The Gospel as Public Truth*. Grand Rapids: Eerdmans, 1991.

Noll, Mark A. *Jesus Christ and the Life of the Mind*. Grand Rapids: Eerdmans, 2011.

———. *The Scandal of the Evangelical Mind*. Grand Rapids: Eerdmans, 1994.

BIBLIOGRAPHY

O'Day, Gail R. "John." In vol. 9 of *New Interpreter's Bible Commentary*, edited by Leander E. Keck. Nashville: Abingdon, 1995.

Oord, Thomas Jay, ed. *Creation Made Free: Open Theology Engaging Science*. Eugene, OR: Wipf & Stock, 2009.

———. *The Uncontrolling Love of God: An Open and Relational Account of Providence*. Downers Grove: InterVarsity, 2015.

Pannenberg, Wolfhart. *Towards a Theology of Nature: Essays on Science and Faith*. Edited by Ted Peters. Louisville: Westminster John Knox, 1993.

Pascal, Blaise. *The Mind on Fire: A Faith for the Skeptical and Indifferent*. Abridged and edited by James M. Houston. Vancouver: Regent College Publishing, 2003.

Peacocke, Arthur. *Theology for a Scientific Age: Being and Becoming—Natural, Divine and Human*. Minneapolis: Fortress, 1993.

Peterson, Michael, et al. *Reason and Religious Belief: An Introduction to the Philosophy of Religion*. 4th ed. Oxford: Oxford University Press, 2009.

Pinnock, Clark H. *Most Moved Mover: A Theology of God's Openness*. Grand Rapids: Baker, 2001.

Pinnock, Clark H., et al. *The Openness of God: A Biblical Challenge to the Traditional Understanding of God*. Downers Grove: InterVarsity, 1994.

Plantinga, Alvin. *Where the Conflict Really Lies: Science, Religion, and Naturalism*. Oxford: Oxford University Press, 2011.

Plantinga, Richard J., et al. *An Introduction to Christian Theology*. Cambridge: Cambridge University Press, 2010.

Polkinghorne, John. *Exploring Reality: The Intertwining of Science and Religion*. New Haven: Yale University Press, 2005.

———. *The Faith of a Physicist: Reflections of a Bottom-Up Thinker*. Minneapolis: Fortress, 1996.

———. *Faith, Science and Understanding*. London: SPCK, 2000.

———. *The God of Hope and the End of the World*. New Haven: Yale University Press, 2002.

———. *One World: The Interaction of Science and Theology*. London: SPCK, 1986.

———. *Quantum Physics and Theology: An Unexpected Kinship*. New Haven: Yale University Press, 2007.

———. *Reason and Reality: The Relationship between Science and Theology*. London: SPCK, 1991.

———. *Science and Creation: The Search for Understanding*. London: SPCK, 1988.

———. *Science and Providence: God's Interaction with the World*. West Conshohocken: Templeton, 2005.

———. *Testing Scripture: A Scientist Explores the Bible*. Grand Rapids: Brazos, 2010.

———, ed. *The Work of Love: Creation as Kenosis*. Grand Rapids: Eerdmans, 2001.

Polkinghorne, John, and Nicholas Beale. *Questions of Truth: Fifty-One Responses to Questions about God, Science and Belief*. Louisville: Westminster John Knox, 2009.

Popper, Karl. *The Logic of Scientific Discovery*. New York: Routledge, 2002.

Reddish, Tim. "The Dawn." In *The Amish Farmer Who Hated L.A.: And 8 Other Modern-Day Allegories*, 15–26. Sisters, OR: Deep River, 2015.

Rees, Martin J. *Just Six Numbers: The Deep Forces That Shape the Universe*. London: Phoenix Press, 2000.

Rice, Richard. "Process Theism and the Open View of God: The Crucial Difference." In *Searching for an Adequate God: A Dialogue between Process and Free Will Theists*,

Bibliography

edited by John B. Cobb Jr. and Clark H. Pinnock, 163–200. Grand Rapids: Eerdmans, 2000.

———. *Suffering and the Search for Meaning: Contemporary Responses to the Problem of Pain*. Downers Grove: InterVarsity, 2014.

Ross, Hugh. *The Creator and the Cosmos: How the Greatest Scientific Discoveries of the Century Reveal God*. 2nd ed. Colorado Springs: Navpress, 1995.

Russell, Robert John. *Cosmology from Alpha to Omega: The Creative Mutual Interaction of Theology and Science*. Minneapolis: Fortress, 2008.

Sanders, John. *The God Who Risks: A Theology of Divine Providence*. 2nd ed. Downers Grove: InterVarsity, 2007.

Schaff, Philip. *The Creeds of Christendom*. Vol 3. http://www.ccel.org/ccel/schaff/creeds3.

Silva, Moisés. "Contemporary Theories of Biblical Interpretation." In vol. 1 of *New Interpreter's Bible Commentary*, edited by Leander E. Keck, 107–24. Nashville: Abingdon, 1994.

Squires, Euan. *Conscious Mind in the Physical World*. Bristol: Hilger, 1990.

———. *The Mystery of the Quantum World*. Bristol: Hilger, 1986.

Stanesby, Derek. *Science, Reason and Religion*. London: Routledge, 1985.

Stannard, Russell. *Science and Belief: The Big Issues*. Oxford: Lion Hudson, 2012.

Swinburne, Richard. *The Concept of Miracle*. Basingstoke: Macmillan, 1970.

Tiessen, Terrance. *Providence and Prayer: How Does God Work in the World?* Downers Grove: InterVarsity, 2000.

Toulmin, Stephen. *The Return to Cosmology*. Berkeley: University of California Press, 1982.

Van Der Toorn, Karel. "Baal." In *New Interpreter's Dictionary of the Bible*, 1:367–69. Nashville: Abingdon, 2006.

Van Till, Howard J. *The Fourth Day: What the Bible and the Heavens Are Telling Us about the Creation*. Grand Rapids: Eerdmans, 1986.

Wall, Robert. W. "The Canonical View." In *Biblical Hermeneutics: Five Views*, edited by Stanley E. Porter and Beth M. Stovell, 111–30, 188–200. Downers Grove: InterVarsity, 2012.

Walls, Jerry L., and Joseph R. Dongell. *Why I Am Not a Calvinist*. Downers Grove: InterVarsity, 2004.

Walton, John H. *The Lost World of Genesis One: Ancient Cosmology and the Origins Debate*. Downers Grove: InterVarsity, 2009.

Ward, Keith. *The Big Questions in Science and Religion*. West Conshohocken: Templeton, 2008.

———. *Divine Action: Examining God's Role in an Open and Emergent Universe*. West Conshohocken: Templeton, 2007.

———. *God, Chance and Necessity*. Oxford: Oneworld, 1996.

Ware, Bruce A., ed. *Perspectives on the Doctrine of God: Four Views*. Nashville: Broadman & Holman, 2008.

Westminster Confession. http://www.bible-researcher.com/wescontext.html.

Westphal, Merold. "The Philosophical/Theological View." In *Biblical Hermeneutics: Five Views*, edited by Stanley E. Porter and Beth M. Stovell, 70–88, 160–73. Downers Grove: InterVarsity, 2012.

———. *Whose Community? Which Interpretation? Philosophical Hermeneutics for the Church*. Grand Rapids: Baker, 2009.

Bibliography

White, Lynn, Jr. "The Historical Roots of Our Ecological Crisis." *Science* 155 (1967) 1203–7.

Whitehead, Alfred North. *Modes of Thought*. New York: Simon & Schuster, 1968.

Wilkinson, David. "Reading Genesis 1–3 in the Light of Modern Science." In *Reading Genesis After Darwin*, edited by Stephen C. Barton and David Wilkinson, 127–44. Oxford: Oxford University Press, 2009.

———. *When I Pray What Does God Do?* Oxford: Monarch, 2015.

Willard, Dallas. *The Divine Conspiracy: Rediscovering Our Hidden Life in God*. New York: HarperOne, 1997.

Winslow, Karen Strand. "The Earth Is Not a Planet." In *Creation Made Free: Open Theology Engaging Science*, edited by Thomas Jay Oord, 13–27. Eugene, OR: Wipf & Stock, 2009.

Witherington, Ben, III. *New Testament Rhetoric: An Introductory Guide to the Art of Persuasion in and of the New Testament*. Eugene, OR: Cascade, 2009.

Wolterstorff, Nicholas. "Unqualified Divine Temporality." In *God and Time: Four Views*, edited by Gregory E. Ganssle, 187–213. Carlisle: Paternoster, 2002.

Wright, J. E. "Cosmogony, Cosmology." In *New Interpreter's Dictionary of the Bible*, 1:755–63. Nashville: Abingdon, 2006.

Wright, N. T. *Evil and the Justice of God*. Downers Grove: InterVarsity, 2006.

———. *Jesus and the Victory of God*. Minneapolis: Fortress, 1996.

———. *Surprised by Hope: Rethinking Heaven, the Resurrection, and the Mission of the Church*. New York: HarperOne, 2008.

Ziman John M. *The Force of Knowledge: The Scientific Dimension of Society*. Cambridge: Cambridge University Press, 1976.

General Index

accommodation, divine, 6–8, 12, 14, 28, 41, 65
action, divine, 40, 65, 78, 86–87, 89, 99, 103–9, 113, 123, 130, 133, 138, 142–43, 148
anthropic principle, 72–76, 79
apologetics, apologist, 5, 79–81, 136
Apophis, 151–52
argument, cosmological, 69, 72, 171–73
argument, first cause; see: argument, Cosmological
argument, from design; see: argument, Teleological
argument, ontological, 171
argument, teleological, 72–80, 171, 173–74
Aristotelianism, 3–6, 12, 19–20, 49, 54–55, 66, 103
authorial intent, 30–32, 38
authority, biblical: "bottom-up," 38–40
authority, biblical: "top-down," 37, 39

Baal, 156–57
Babel, Tower of, 84
Behemoth, 162
biblicism, 177
big bang, 21, 72–73, 82, 172

canon, of Scripture, 9, 24–27, 31, 37, 39, 150, 166–67
canonical criticism, 39–40, 167
causation, "top-down," 105
cause, primary, 85–86, 103
cause, secondary, 85–86

chance, xv, 74–75, 77, 82, 87–88, 92–108, 127, 160, 162, 173–74
chaos, 62, 68, 79, 95, 103, 107, 147, 151–53, 155–66
closed system, 45, 111, 121, 129, 145; see also: universe, clockwork
complexity, irreducible, 62, 76–77
concordism, 14, 80–82, 145
conflict, science-faith, xiv, 2, 10, 15, 19, 40, 43, 61–65, 67–68, 78, 90, 103, 108, 144, 177
contingency, 69, 89, 107–8, 172–73
control, divine, xv, 103–7, 113–14, 120, 122, 162
Council of Trent, 2, 8–10, 13
counterfactuals, 119–21
creation-out-of-nothing, or *creatio ex nihilo*, 88, 95, 112, 163–65
creation-out-of-chaos, or *creatio ex materia*, 88, 163–65
creation, continuous, 66, 84, 86, 99, 121, 128, 137, 150, 162, 164–66
creation, (Genesis 1) 6, 21, 61, 80–81, 83–84, 145–50, 155–57, 162, 164–66
creation, (Genesis 2–3) 83–84, 145, 150–56, 164–66
crisis, of significance, 33

death of the author, 32, 80
death of God, 32
death, physical, 67, 79, 100, 151–55, 166
deism, deist, 78–79, 85–87, 98, 103, 105, 110–11, 125

187

General Index

demiurge, 110
determinism, divine, 103–8, 114, 117–119, 121, 140; see also control, divine
determinism, naturalistic, 54, 56–57, 78, 86, 99–100, 103–5, 107–8, 129, 131
dialogue, science-faith, xi, xiii–xiv, 18, 41, 61, 68–71, 80, 82, 85, 90, 103, 108–10, 126–27, 140, 145, 167, 169–70
dualism, body-mind, 67
dualism, body-soul, 67
dualism, subject-object, 34

ecological crisis, or environmental crisis, 83–84, 148–50
eisegesis, 38, 71, 92, 145, 148
Enuma Elish, 156–57
eschatology, or eschaton, 39, 67, 84, 89, 133, 135, 160, 165
everlastingness, of God, 115–16, 118, 123
evil, 67, 79, 82, 86, 88, 95, 98, 105, 107, 111, 114, 119–20, 122, 125, 128, 135, 137, 148, 151–52, 154, 158, 162, 165
evolution, biological, xii–xiii, 16, 19–21, 61, 63, 66, 75–80, 85–87, 89, 93, 96–98, 104, 107, 122, 145, 174–75, 178
evolution, cosmic, 21, 74, 80, 174
exegesis, 3, 14–15, 34, 38, 83, 167
existentialism, existential, 34–35, 58, 65, 72, 90

fall, of Adam and Eve, 8, 84, 153–55, 166
falsificationism, falsification, 50–56, 130, 177
flood, Noah's, 17, 155, 161
foreknowledge, simple, 117–19
free process defense, 98–99, 107
free will, 25, 40, 88, 98, 105, 113, 116–21, 142

garden of Eden, 150–51, 155
geocentrism, geocentric, 11, 15, 20–21

Gilgamesh epic, 151–52
God-of-the-gaps, 45, 75, 77–78
god of the philosophers, 123, 171

heliocentrism, heliocentric, xiv, 1–2, 9–15, 18–22, 55
hermeneutical circle, 35–36, 40, 50
hermeneutics, xii, xiv, 3, 17, 27–40, 58, 81–82, 103, 106
hidden variables, 100, 104
historical criticism, 29–34, 37, 39–40, 50, 81–82

idealism, 176
ideals, 44
image of God, or *imago Dei*, 20, 83, 85, 149, 155
immanence, divine, 66, 68, 72, 84, 110, 123
immutability, Aristotelian, 12, 54
immutability, divine, 90, 103, 111–12, 115, 118, 123, 140
impassibility, divine, 90, 111–12
independence, science-faith, xiv, 5, 14, 16, 61, 64–68, 72, 85, 90
indeterminacy, quantum, xv, 69, 93, 99–108, 172
induction, scientific, 46, 48, 50, 52–55, 94, 129–30
inerrancy, biblical, 6, 10, 14, 17, 25, 37
infallibility, biblical, see: inerrancy, biblical
inspiration, biblical, xii, xiv, 7, 10, 23–28, 32, 37–41, 64, 106, 109, 117
instrumentalism, 101, 175–78
integration, science-faith, xiv, 61, 68, 71–90, 103, 108, 167
intelligent design, 72, 76–80
interpretation, biblical, xii–xiv, 3–15, 17, 19, 22–41, 59, 61, 64, 80–81, 117

kenosis, or divine self-limitation, 98, 121
kingdom of God, 132, 134–35, 137, 160

General Index

law, scientific, 49, 52, 85, 87–89, 93–100, 103–8, 127–32, 155
Laplace, 86, 99–100, 103, 140
Leviathan, 155, 158–60, 162, 165
literalism, biblical, 7, 9, 17, 62–64, 145

many-universes hypothesis, 102
materialism, scientific, 61–64, 104, 177–78
metanarrative, 33, 39
middle knowledge, 119–20
miracle, xv, 67, 84, 88, 90, 94, 98, 105, 125–38, 142
modernism, 17–18, 33, 62, 79, 108, 145
Molinism; see: middle knowledge
multiverse hypothesis, 74–76, 102

natural theology, 65, 71–82, 88, 95, 104, 171–74
nature, theology of, 71–72, 80, 82–85, 90, 95, 104
necessity, 82, 87, 89, 97–98, 107–8, 172; see also law, scientific
nihilism, 33

observation, scientific, 12, 14, 16, 19, 43–44, 46–55, 58–59, 99–102, 104, 132, 176–77
omnipotence, xv, 85, 88, 90, 103, 111–14, 118, 121, 164–65
omnipresent, 115, 121
omniscience, 90, 102–3, 106, 111, 115–23, 139
open system, 89, 97–98, 102, 108, 111, 120–21, 125, 133, 142, 149, 163

panentheism, 84, 88
paradigm, 28–29, 34, 38–39, 54–56, 58, 62, 71, 90, 94, 101, 103, 111, 176–77
persuasion, divine, 87–89, 107–8; see also: process theology
philosophy, natural, 2–6, 10, 14–15, 40, 60
physics, classical, or Newtonian, 34, 54–57, 100–101, 103, 111

postmodernism, postmodern, xiii–xiv, 18, 33–34, 37–38, 42, 68, 79, 108, 170
prayer, xv, 78, 84, 98, 105, 118, 125–28, 133, 138–43
predestination, predestine, 92, 98, 102–5, 108, 117, 119, 140
process theology, 85, 87–90, 109, 122, 132, 164

quantum mechanics, or quantum world, xv, 34, 54–56, 93–95, 99–108, 172, 176–77

realism, 101, 175–77
realism, critical, 177
reductionism, 57, 62–63, 104, 128

salvation, 10, 12–14, 17, 40–41, 71, 92, 114
scientific method, 28, 43, 46–50, 58, 175
Scripture, fuller sense, 28, 31, 80
Scripture, the purpose of, xiv, 14, 17, 40–41, 81
sea monsters, sea, 144, 147, 155–62, 166; see also: Leviathan
self-organization, 62, 69, 86, 96–98
sense, allegorical, 3–4, 6, 28
sense, anagogical, 4
sense, literal, 3–4, 6, 9, 11, 17, 25, 27–28, 61, 65, 80, 126–27, 146, 158, 166; see also: literalism, biblical
sense, tropological, 4
serpent, or snake, 151–53, 158, 160
shalom, 90, 134–35, 153
sola scriptura, xiv, 9, 11, 117, 167
sovereignty, divine, 85, 103, 106, 113–14, 118, 120, 122, 160, 165
suffering, 79, 85–86, 89, 98, 112, 119, 133, 140, 161–63
supernaturalism, supernatural, 77, 88, 105, 126, 128, 131–32, 134, 137, 151
synthesis, systematic, 71, 85–90

theism, classical, 88, 118–20, 123, 140
theism, open, 120, 122

theism, relational, 120, 122
theodicy, 86, 105
Thomism, 85–86, 117
Tiamat, 151, 156, 158
time, and God, xv, 84, 102–3, 106, 110–11, 115–23, 133, 139–40, 143
timelessness, of God, 116, 118, 123
tradition, two books, 90, 93
transcendence, divine, 65–66, 68, 72, 85, 88–89, 110–11, 122–23
Tree of Knowledge, 151–52
Tree of Life, 150–53
Trinity, Trinitarian, 71, 88, 109–12, 114, 116, 133, 138, 148, 165
truth, private, 59, 66

truth, public, 59, 66

uncertainty principle, Heisenberg, 100–102, 104, 176, 178; see also: indeterminacy, quantum
universe, clockwork, or mechanistic, xv, 54, 56, 78, 86, 93, 98–99, 102–3, 111, 129–30, 136, 141; see also: determinism, naturalistic

verification, scientific, 46, 130

wave-particle duality, 99, 101

Yam, 156–58